Eliza Ann Youmans

The first book of botany

Designed to cultivate the observing powers of children

Eliza Ann Youmans

The first book of botany
Designed to cultivate the observing powers of children

ISBN/EAN: 9783337041694

Printed in Europe, USA, Canada, Australia, Japan

Cover: Foto ©ninafisch / pixelio.de

More available books at **www.hansebooks.com**

THE

FIRST BOOK OF BOTANY.

DESIGNED TO CULTIVATE

THE OBSERVING POWERS OF CHILDREN.

.... Not that *more* is taught at an early age, but *less*; that time is taken; that the wall is not run up in haste; that the bricks are set on carefully, and the mortar allowed time to dry.

LORD STANLEY.

BY

ELIZA A. YOUMANS.

NEW YORK:
D. APPLETON AND COMPANY,
90, 92 & 94 GRAND STREET.
1870.

ENTERED according to Act of Congress, in the year 1870, by
D. APPLETON AND COMPANY,
In the Clerk's Office of the District Court of the United States for the Southern District of New York.

PREFACE.

This little book has a twofold claim upon those concerned in the work of education.

In the first place, it introduces the beginner to the study of Botany in the only way it can be properly done—by the direct observation of vegetable forms. The pupil is told very little, and from the beginning, throughout, he is sent to the plant to get his knowledge of the plant. The book is designed to help him in this work, never to supersede it. Instead of memorizing the statements of others, he brings report of the living reality as he sees it; it is the things themselves that are to be examined, questioned, and understood. The true basis of a knowledge of Botany is that familiarity with the actual characters of plants, which can only be obtained by direct and habitual inspection of them. The beginner should therefore commence with the actual specimens, and learn to distinguish those external characters which lie open to observation; the knowledge of which leads naturally to that arrangement by related attributes which constitutes classification.

But the plan of the work claims the attention of those interested in education, on other and more important grounds. Valuable as may be a knowledge of the vegetable kingdom, I should hardly have undertaken to make a school-book with reference to this object alone. It is not what Botany is, considered in itself, but what it is capable of doing for the minds of those who pursue it aright, that gives it its highest interest to the educator; and it has been to secure certain important results in mental cultivation which are but imperfectly provided for in our system of popular education, that has led to the preparation of this series of exercises. It is because Botany, beyond all other subjects, is suited to maintain the mind in direct intercourse with the objects and order of Nature, and to train the observing powers and the mental operations they involve, in a systematic way, that I have undertaken to put its rudiments into such a shape that this desirable work can be rightly commenced.

A preface is not the appropriate place to deal with so large a subject as is here touched upon, and which involves nothing less than the true mental philosophy of education; but so important is the question, in its practical bearings, that I have entered into a somewhat full discussion of it at the close of the volume, and would earnestly urge not only parents and teachers, but the responsible directors of public education, to weigh carefully the considerations there presented.

It is needful here to state that the method of instruction developed in these pages is no mere educational novelty; it has been tested, and its fitness for the end proposed has been shown in practice. The *schedule* feature which is here fully brought out, and which is its leading peculiarity as a mode of study, was devised and successfully used by Prof. J. S. Henslow, of Cambridge, England. My attention was first drawn to it as I was looking about in the educational department of the South Kensington Museum, in London. In a show-case of botanical specimens, I noticed some slates covered with childish handwriting, which proved to be illustrations of a method of teaching Botany to the young. They were furnished by Prof. Henslow for the International Exhibition of 1851. He died without publishing his method, but not without having subjected it to thorough practical trial. He had gathered together a class of poor country children, in the parish where he officiated as clergyman, and taught them Botany by a plan similar to the present, though less simplified. The results of this experiment have been given to the public by Dr. J. D. Hooker, Superintendent of the Botanical Gardens at Kew, who was summoned to give evidence upon the subject before a Parliamentary Commission on Education.

The following interesting passages from his testimony will give an idea of Prof. Henslow's method of proceeding and its results:

Question. Have you ever turned your attention at all to the possibility of teaching Botany to boys in classes at school?

Answer. I have thought that it might be done very easily; that this deficiency might be easily remedied.

Q. What are your ideas on the subject?

A. My own ideas are chiefly drawn from the experience of my father-in-law, the late Prof. Henslow, Professor of Botany at Cambridge. He introduced Botany into one of the lowest possible class of schools—that of village laborers' children in a remote part of Suffolk.

Q. Perhaps you will have the goodness to tell us the system he pursued?

A. It was an entirely voluntary system. He offered to enroll the school children in a class to be taught Botany once a week. The number of children in the class was limited, I think, to forty-two. As his parish contained only one thousand inhabitants, there never were, I suppose, the full forty-two children in the class; their ages varied from about eight years old to about fourteen or fifteen. The class mostly consisted of girls. . . . He required that, before they were enrolled in the class, they should be able to spell a few elementary botanical terms, including some of the most difficult to spell, and those that were the most essential to begin with. Those who brought proof that they could do this were put into the third class; then they were taught once a week, by himself generally, for an hour or an hour and a half, sometimes for two hours (for they were exceedingly fond of it).

Q. Did he use to take them out in the country, or was it simply lessons in the school?

A. He left them to collect for themselves; but he visited his parish daily, when the children used to come up to him, and bring the plants they had collected; so that the lessons went on all the week round. There was only one day in the week on which definite instruction was given to the class; but on Sunday afternoon he used to allow the senior class, and those who got marks at the examinations, to attend at his house. . . .

Q. Did he find any difficulty in teaching this subject in class?

A. None whatever; less than he would have had in dealing with almost any other subject.

Q. Do you know in what way he taught it? did he illustrate it?

A. Invariably; he made it practical. He made it an objective study. The children were taught to know the plants, and to pull them to pieces; to give their proper names to the parts; to indicate the relations of the parts to one another; and to find out the relation of one plant to another by the knowledge thus obtained.

Q. They were children, you say, generally from eight to twelve?

A. Yes, and up to fourteen.

Q. And they learned it readily?

A. Readily and voluntarily, entirely.

Q. And were interested in it?

A. Extremely interested in it. They were exceedingly fond of it.

Q. Do you happen to know whether Prof. Henslow thought that the study of Botany developed the faculties of the mind —that it taught these children to think? and do you know whether he perceived any improvement in their mental faculties from that?

A. Yes; he used to think it was the most important agent that could be employed for cultivating their faculties of observation, and for strengthening their reasoning powers.

Q. He really thought that he had arrived at a practical result?

A. Undoubtedly; and so did every one who visited the school or the parish?

Q. They were children of quite the lower class?

A. The laboring agricultural class.

Q. And in other branches receiving the most elementary instruction?

A. Yes.

Q. And Prof. Henslow thought that their minds were more developed; that they were become more reasoning beings, from having this study superadded to the others?

A. Most decidedly. It was also the opinion of some of the inspectors of schools, who came to visit him, that such children were in general more intelligent than those of other parishes; and they attribute the difference to their observant and reasoning faculties being thus developed. . . .

Q. So that the intellectual success of this objective study was beyond question?

A. Beyond question. . . . In conducting the examinations of medical men for the army, which I have now conducted for several years, and those for the East-India Company's Service, which I have conducted for, I think, seven years, the questions which I am in the habit of putting, and which are *not* answered by the majority of the candidates, are what would have been answered by the children in Prof. Henslow's village-school. I believe the chief reason to be, that these students' observing faculties, as children, had never been trained—such faculties having lain dormant with those who naturally possessed them in a high degree; and having never been developed, by training, in those who possessed them in a low degree. In most medical schools, the whole sum and substance of botanical science is crammed into a few weeks of lectures, and the men leave the class without having acquired an accurate knowledge of the merest elements of the science. . . .

The printed form or *schedule* contrived by Prof. Henslow, and used in these classes, applied only to the flower, the most complex part of the plant, and the attention of children was directed by it chiefly to those features upon which orders depend in classification. But, instead of confining its use to the study of a special part of plant-structure, it seemed to me to apply equally to the whole course of descriptive Botany, and to be capable of becoming a most efficient instrument of regular observational training. I accordingly prepared a simplified series of exercises on this plan, and used them to guide some little children in studying the plants of the neighborhood; and, had this experiment not been regarded, by those who witnessed it, as a success, the book embodying these exercises would not now appear.

As the plan of teaching here adopted is to tell the child as little as possible, thereby limiting the text to bare definitions, to be employed in connection with the pictures, it happened that there was not sufficient matter to fill up the vacancies between the numerous cuts. But, rather than deviate from this plan, I preferred to occupy the spaces with hints to teachers, which will account for the brief notes interspersed through the book.

Should this First Book prove acceptable, a second may follow it upon a similar plan, for the use of those who wish to go on with the subject by a similar method.

Although designed for beginners, and presenting only the barest rudiments of the subject, yet the preparation of this little book for the end I had in view, and under the test of practical trial, has not been a trifling task; but, if only a small number of the young shall be led by it to a familiar acquaintance with their interesting little "neighbors by the wayside," and shall thereby learn to look with their own eyes, and to think with their own minds, I shall be well repaid for the solicitude it has cost.

I am much indebted to my brother, E. L. Youmans, for his kind assistance in taking it through the press, and especially in the preparation of the argument on the Educational Claims of Botany, at the close of the volume.

NEW YORK, *February*, 1870.

SUGGESTIONS TO TEACHERS.

THE method to be pursued by the aid of this book is the following: The child, whether at home or at school, first of all collects some specimens of plants—almost any will answer the purpose in commencing. These consist of organs, each of which is made up of different parts, and these vary in form and structure continually in different species. The object of the learner is to find out these parts or characters, and to learn their names, so as to be able to describe them.

The beginner, of course, must start with the simplest characters. Turning to the first exercises, for example, he finds the parts of leaves represented by pictures accompanied by the names applied to them. Guided by these, he refers to his specimens, and *finds the real things* which the pictures and the words represent. When a few characters are fixed in the mind by two or three exercises, he will commence the practice of *noting down* what he observes. For this purpose a form, or *schedule*, is used, containing questions which indicate what he is to search for. Models of these schedules, filled out, are given in the successive exercises: the pupil will make them for himself with pencil and paper.* He now carefully observes his specimen, and writes down the characters it possesses, with which he has thus far become acquainted. Having done this, he pins the specimen to the paper describing it, and brings it to the teacher as the report of his observation and judgment in the case.

---

* I have thought it desirable also to present the whole set, at the end of the volume, with the answers omitted, to illustrate at a glance the scope of this first series of observations. As the pupil is to be constantly engaged in *schedule practice*, and as the schedules are not to be preserved, the cheapest kind of paper will answer, and it can be of course used on both sides. Slates will do just as well; but then the description must be numbered, and a corresponding number attached to the specimen, so that they can be compared by the teacher.

This operation is constantly repeated upon varying forms, and slowly extended by the addition of new characters. He thus goes on discovering new parts and acquiring their names —noting the variations of these parts and the names of their variations. The schedules guide him forward in the right direction, and hold him steadily to the essential work of exercising his faculties upon the living objects before him. In every fresh collection of plants, new parts and new relations will solicit the attention, and will have to be observed, compared, and recorded. Particular kinds of plants, let it be remembered, *are not described in the book*—they are not even named; the object is, by constant practice and repetition, to train the pupil to find out the characters of any that come in his way, and make his own descriptions.

An acquaintance with Botany, although of course desirable, is not indispensable in using these exercises. Any teacher or parent *who is willing to take the necessary pains* can conduct the children through them without difficulty; and if they will become *fellow-students* with them all the better. The child is not so much to be *taught*, as to instruct himself. The very essence of the plan is, that he is to *make his own way*, and rely on nobody else; it is intended for self-development. Mistakes will, of course, be made; but the whole method is *self-correcting*, and the pupil, as he goes forward, will be constantly rectifying his past errors. The object is less to get perfect results at first than to get the pupil's opinion upon the basis of his own observations.

Children can begin to study plants successfully by this method at six or seven years of age, or as soon as they can write. But close observations should not be required from young beginners, nor the exercises be prolonged to weariness. The transition from the unconscious and spontaneous observations of children to conscious observation with a definite purpose should be gradual, beginning and continuing for some time with the easiest exercises upon the most simple and obvious characters.

CONTENTS.

	PAGE
CHAPTER I.—THE LEAF	15
Ex. 1. The Parts of a Leaf	16
2. The Parts of a Grass-Leaf	17
3. Venation	18
4. The Framework and its Parts	19
5. Feather-veined and Palmate-veined Leaves	22
6. Margins	24
7. Bases	28
8. Apexes	31
9. Forms of Lobes	33
10. Forms of Sinuses	35
11. Kinds of Leaves	37
12. Shapes of Leaves	39
13. Petioles, Surfaces, and Colors	45
14. Simple and Compound Leaves	47
15. Parts of Compound Leaves	49
16. Pinnate and Digitate Leaves	51
17. Varieties of Pinnate Leaves	52
18. Varieties of Digitate Leaves	54
19. Forms of Stipules	57
20. Examples of Description	59
CHAP. II.—THE STEM	60
Ex. 21. Parts of the Stem, and Leaf Axil	60
22. Appendages of the Stem	62
23. Position of Leaves	64
24. Arrangement of Leaves on the Stem	66
25. Shapes of Stems	69
26. Attitude of Stems	71
27. Color, Surface, Size, Structure	74

	PAGE
CHAP. III.—THE INFLORESCENCE...	76
Ex. 28. Solitary and Clustered Inflorescence...	76
29. Parts of the Inflorescence...	78
30. Attitude of Inflorescence...	80
31. Solitary Terminal and Axial Inflorescence...	82
32. Clustered Axial and Terminal Inflorescence...	84
33. Definite and Indefinite Inflorescence...	87
34. Varieties of Inflorescence...	90
CHAP. IV.—THE FLOWER...	96
Ex. 35. Parts of the Flower...	96
36. Parts of the Calyx...	97
37. Parts of the Corolla...	98
38. Kinds of Calyx...	99
39. Kinds of Corolla and Perianth...	100
40. Regular and Irregular Corollas and Perianths...	101
41. Parts of Stamens...	103
42. Parts of the Pistil...	105
43. Parts of the Ovary...	105
44. Parts of the Petals...	107
45. Kinds of Regular Polypetalous Corollas...	108
46. Kinds of Irregular Polypetalous Corolla...	109
47. Parts of a Gamopetalous Corolla...	112
48. Kinds of Regular Gamopetalous Corollas...	113
49. Irregular Gamopetalous Corollas...	114
50. Crowns, Spurs, and Nectaries...	118
CHAP. V.—THE ROOT...	120
Ex. 51. Tap-Roots and Fibrous Roots...	120
52. Kinds of Tap-Root...	121
53. Kinds of Fibrous Roots...	122
EXAMPLES IN PLANT DESCRIPTION...	124
LEAF SCHEDULES...	143
STEM SCHEDULES...	148
INFLORESCENCE SCHEDULES...	151
FLOWER SCHEDULES...	154
THE EDUCATIONAL CLAIMS OF BOTANY...	158

The First Book of Botany

CHAPTER I.—THE LEAF.

THE pupil will see from the picture what is to be done first, and how we are to proceed in commencing the study of plants. Having collected some specimens, let us begin with the leaf. On these printed leaves there is a language which children have already learned; there is also a language written by Nature on the *leaves that grow:* we will now learn to read *that*.

EXERCISE I.

The Parts of a Leaf.

The beginner will gather some leaves, and find out the names of their parts by comparing them with the picture.

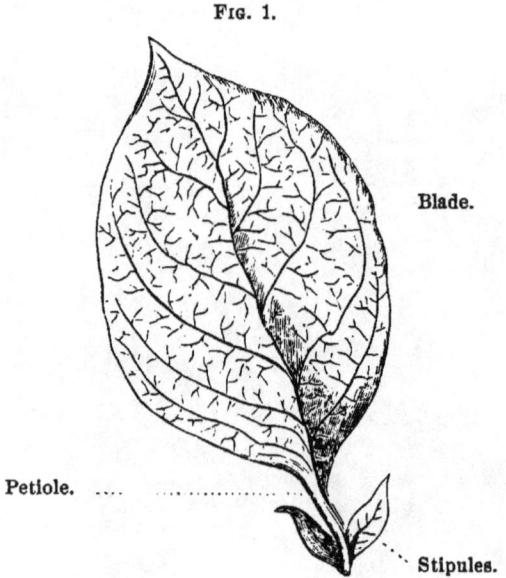

Fig. 1.

BLADE.—The flattened green part of the leaf.

PET'IOLE.—The leaf-stem.

STIP'ULES.—Small bodies at the base of the petiole, that look more or less like leaves.

NOTES FOR TEACHERS.—The exercises begin with leaves, because they are the simplest and the most common parts of plants, and because they present the greatest variety of forms, and are most easily procured. The aim of the first exercise is to teach the parts of a leaf and their names. It is likely that the first gathering of leaves will be done carelessly, and that,

THE LEAF.

EXERCISE II.

The Parts of a Grass-Leaf.

GATHER a handful of grass and see if you can find the parts shown in Fig. 2.

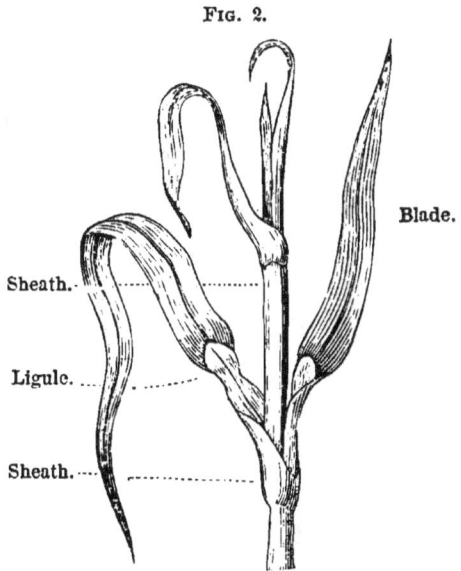

FIG. 2.

BLADE.—The flattened upper part of the leaf.
SHEATH.—A leaf-stem surrounding the stalk.
LIG'ULE.—The scale-like stipule often seen between the sheath and the blade.

---

when compared with Fig. 1, the specimens will be found lacking in some of the parts there seen. This will make it necessary to repeat the exercise. At the second trial the leaves will be pulled with more care, and the pupil will seek for those having all the parts seen in the picture. Let him point out the parts in each of his specimens, and give them their names, repeating the process till he can do it without hesitation or mistake.

EXERCISE III.

Venation.

VENA′TION.—The lines seen upon the leaf-blade are called its venation.

Hold up a leaf between your eye and the light, and, if you see a net-work of lines, like Fig. 3, it is a net-veined leaf; but, if you see no net-work, as in Fig. 4, it is a parallel-veined leaf.

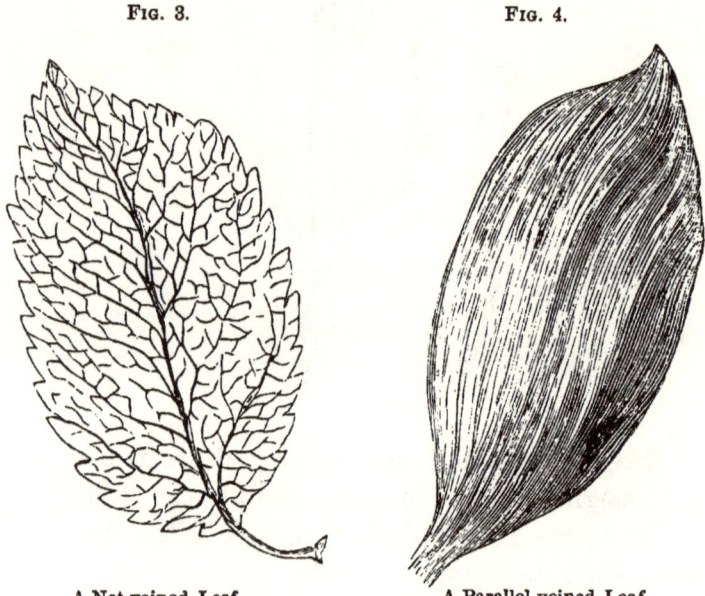

FIG. 3. FIG. 4.

A Net-veined Leaf. A Parallel-veined Leaf.

LOOKING AND OBSERVING.—There are plenty of boys and girls who have always lived in a garden, and yet, if you asked them the difference between a potato-leaf and a bean-leaf, they could not tell you. They have *looked* at potato-plants and bean-plants often enough, but they have never *observed* them.

THE LEAF.

When we observe a thing, we not only look at it, but, as we look, we think particularly about it. For instance, after these exercises, when you look at a leaf, you will think, what parts has it? and, is it net-veined or not? You will *observe* these particulars about it.

THE SCHEDULE.—That you may be sure to look at plants with care, and that your teacher may see what you think about them, little diagrams, called schedules, are used, in which you are to write down what you observe. They have questions written upon them, which you are to answer by studying the plants themselves.

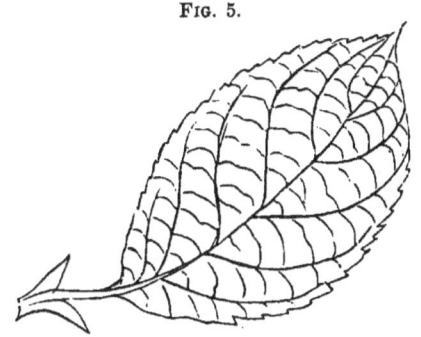

FIG. 5.

SCHEDULE FIRST, DESCRIBING FIG. 5.

Parts?	*Blade, Petiole, Stipules.*
Venation?	*Net-veined.*

NOTE.—It will be observed that the attention of the child is restricted to one additional point at each exercise. This will prevent the confusion of ideas which is liable to arise when several new features of plant-structure are presented to the mind at the same time.

Here is such a schedule about Fig. 5. On the left, two words are printed with interrogation-points, which show that they are questions. The word Parts? means, what parts has this leaf? The word Venation? means, what is its venation? The answers to these questions are found by looking at the picture, and they are then written in the schedule as you see.

Take a sheet of ruled paper, and make a vertical pencil-mark an inch or two from the left edge; at the left of this mark write the questions, Parts? Venation? Now examine a real leaf, and opposite the question, Parts? write what parts you find. Look again at its venation, and write the answer to this question also. Pin each leaf upon the paper that describes it, and hand the collection to the teacher, to see if you have observed correctly.

EXERCISE IV.

The Framework and its Parts.

THE lines upon the blade of a leaf, shown in Fig. 6, are made by its framework. The spaces between

---

NOTE.—A word of caution is here necessary against mistaking the purpose of this book for that of common botanies. The aim of ordinary botanical teaching is simply to impart to pupils a knowledge of plants. In our schools the ambition of both teacher and pupil is to get something done as quickly as possible that will show proficiency. Hence the early attempts at the classification of plants and the consequent precipitation

these lines, which are darkened in Fig. 7, are, in the living leaf, filled with green matter.

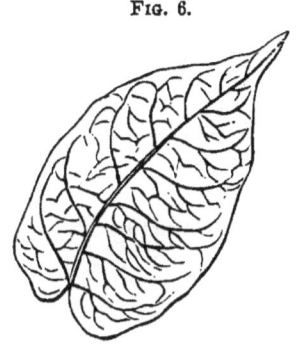

Fig. 6.

Fig. 7.

You know the names of the parts of a *leaf*, and the two following pictures will show you what to call the different parts of the *framework*.

Ribs.—The stoutest pieces of the framework that begin at the petiole and reach quite across the blade, are called *ribs*. When there is but one, as in Fig. 8, it is called a *midrib*.

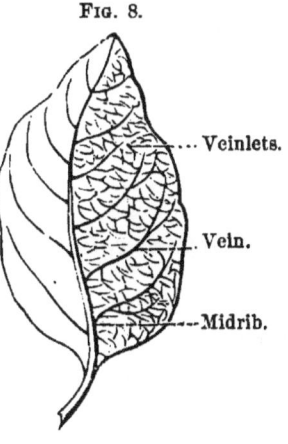

Fig. 8.

... Veinlets.

.. Vein.

---Midrib.

---

of the pupil into the complexities of the subject before the simpler portions have been sufficiently mastered.

Now, the aim of this book is carefully to guard against such a result. These first observations are made without reference to those combinations of characters by which plants are identified as belonging to a particular order, genus, or species. One of our aims is to learn the elementary facts so thoroughly and

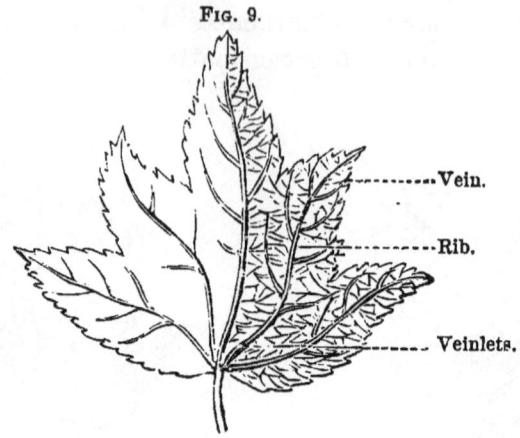

Fig. 9.

V\ESUB{}EINS.—The branches of the ribs are called *veins*.
V\ESUB{}EIN'LETS.—The branches of the veins are called *veinlets*.

EXERCISE V.

Feather-veined and Palmate-veined Leaves.

If you have carefully compared a few living leaves with Figs. 8 and 9, you know the difference between ribs and veins.

familiarly that we may be prepared to go forward and use them afterward. We first study the parts of plants one after another, on account of what they offer directly to observation. When the characters of leaves, stems, flowers, etc., have become familiar, their relations to each other in different plants, which are usually thrust upon the attention at the outset of study, will come to be seen with little effort. This spontaneous action will be sure to occur as soon as the pupil is prepared for it. All that need be done, therefore, is to keep the elements of the subject before the mind, and to acquire the use of accurate

THE LEAF.

Now, when a leaf has but one rib—a midrib—which gives off veins right and left, like Fig. 10, making it look something like a feather, it is called a feather-veined leaf; and when several ribs pass across the blade in a spreading fashion, as in Fig. 11, the leaf is said to be palmate-veined. Whoever named it so, must have thought the ribs looked like the spread-out fingers branching off from the palm of the hand.

FIG. 10.

FIG. 11.

If a leaf is net-veined, it will be in one of these two fashions. It will be either feather-veined or palmate-veined. In answering the question Venation? in your schedule, you may now state whether the leaf in hand is feather-veined or palmate-veined.

You may sometimes be troubled to decide whether a leaf is feather-veined or palmate-veined. Large veins near the base sometimes look very much like *ribs*. Compare your leaf carefully with the pictures and definitions, and write your opinion in the schedule. You may make mistakes at first, but further observation will enable you to correct them.

---

terms in description, without troubling ourselves about the higher growths of the science.

EXERCISE VI.

Margins.

MARGIN.—THE edge of a leaf-blade is called its *margin*.

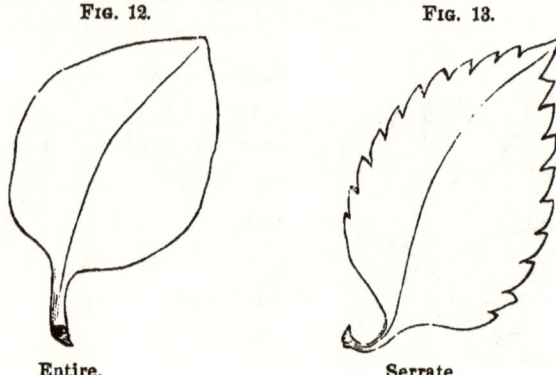

Fig. 12.   Fig. 13.

Entire.   Serrate.

An ENTIRE margin is even and smooth, like Fig. 12.

A SER'RATE margin has sharp teeth pointing forward like a saw (see Fig. 13).

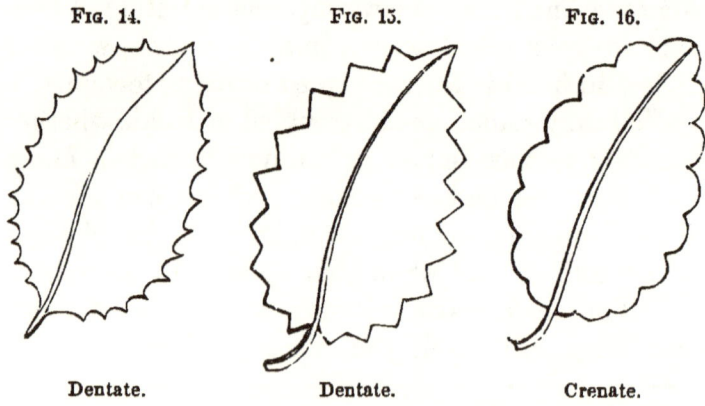

Fig. 14.   Fig. 15.   Fig. 16.

Dentate.   Dentate.   Crenate.

A DEN'TATE margin has sharp teeth pointing out-

THE LEAF.

ward. Figs. 14 and 15 are different forms of Dentate margin.

A CRE'NATE margin has broad, rounded notches, like Fig. 16.

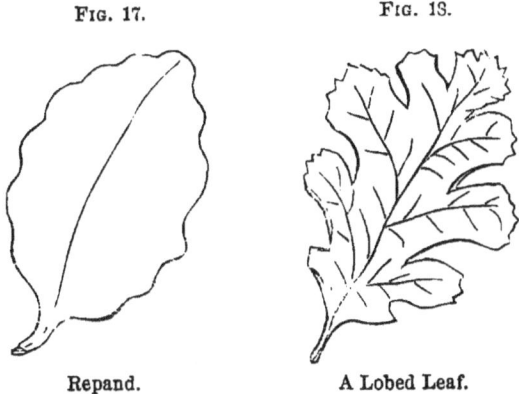

Fig. 17. Fig. 18.

Repand. A Lobed Leaf.

In REPAND' (WAVY) margins the edge curves outward and inward, as in Fig. 17.

Such deep notches as are seen in Fig. 18 form lobes.

Each of these different kinds of margin varies in many ways, and some of the variations are important

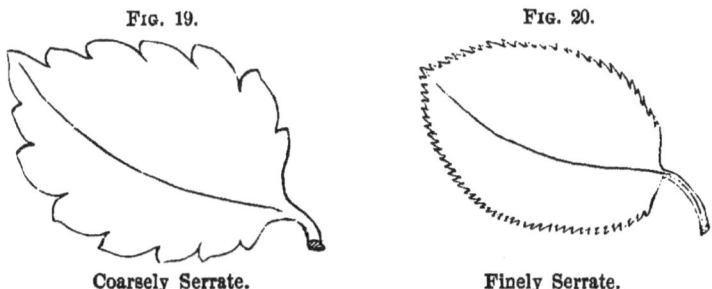

Fig. 19. Fig. 20.

Coarsely Serrate. Finely Serrate.

in description. For instance, serrate margins are sometimes COARSELY SERRATE (Fig. 19), FINELY SERrate (Fig. 20), Doubly Serrate (Fig. 21), and Unevenly Serrate (Fig. 22).

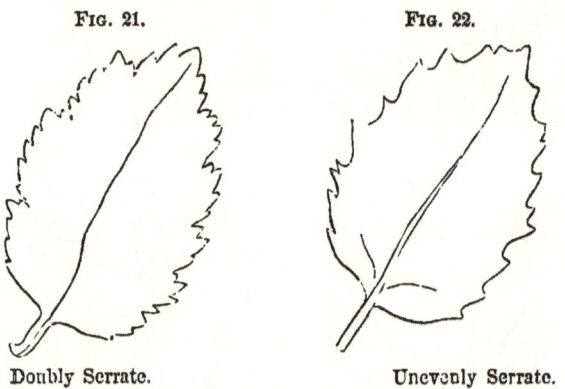

Fig. 21. Doubly Serrate.

Fig. 22. Unevenly Serrate.

Look out for the same kinds of variation among crenate margins. Fig. 23 shows you a Finely Crenate margin. Doubly crenate margins are very common.

Dentate margins are coarse, fine, double, and also uneven.

You will sometimes find two kinds of margin on the same leaf. Part of the notches may be serrate and part dentate, and this forms a serrate-dentate margin. If some of the notches are crenate and some serrate, it will be crenate-serrate, and so on.

In answering the new question, Margin? which you will find in the next schedule, you must look closely for all these different forms, and get familiar with the terms by which they are described.

THE LEAF. 27

Fig. 24.

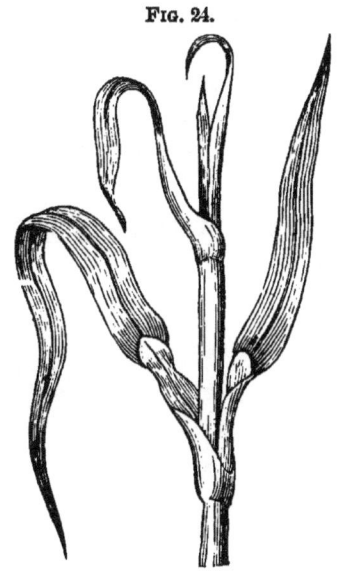

Schedule Second, describing Fig. 24.

Parts?	Blade, Sheath, Ligule.
Venation?	Parallel-veined.
Margin?	Entire.

EXERCISE VII.

Bases.

The Base of a leaf is its lower or attached end. Bases are

Cor'date (Heart-shaped).—Shaped like a heart. Fig. 25.

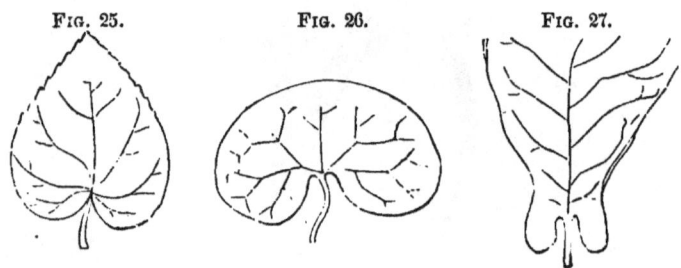

Fig. 25.   Fig. 26.   Fig. 27.

Ren'iform (Kidney-shaped).—Shaped like a kidney. Broader than long. Fig. 26.

Auric'ulate (Ear-shaped).—With small, rounded lobes at the base. Fig. 27.

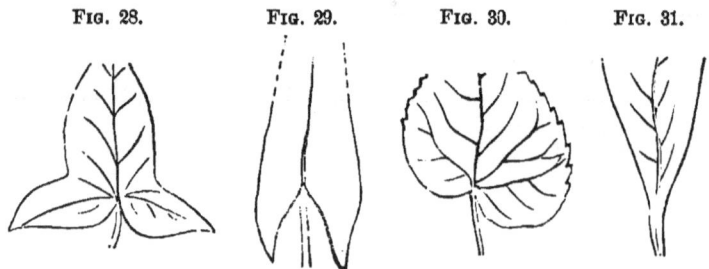

Fig. 28.   Fig. 29.   Fig. 30.   Fig. 31.

Has'tate (Halbert-shaped).—With spreading lobes at the base. Fig. 28.

Sag'ittate (Arrow-shaped).—With sharp lobes at the base pointing backward. Fig. 29.

THE LEAF.   29

OBLIQUE'.—With one side of the base larger and lower than the other. Fig. 30.

TAPERING.—Where the blade tapers off at the base. Fig. 31.

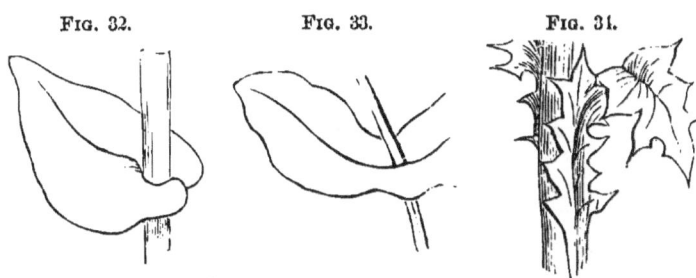

FIG. 32.   FIG. 33.   FIG. 34.

CLASPING.—Where the base folds around the stem of the plant. Fig. 32.

CONNATE'.—Where the bases of two leaves grow together around the plant-stem, as in Fig. 33.

DECUR'RENT.—Where the lower part of the midrib grows to the plant-stem, as in Fig. 34.

NOTE.—Children will, of course, get leaves from the same plants, and describe them over and over again as they pass on from schedule to schedule. A few plants will obtrude themselves upon the attention, and each day the pupil will gather leaves from these alone. At first they will have very little enterprise in searching for new specimens, but will be content with whatever is easiest. These will serve perhaps as well as any to illustrate the new character brought out by the new schedule, but the repetition of old observations upon them will require but little effort of the attention. This repetition of observations upon the same varieties of leaves is proper and desirable, but not sufficient for our purpose. As the wealth of varied forms that plants present is to be our means of educating the observation, it is indispensable that our re-

FIG. 35.

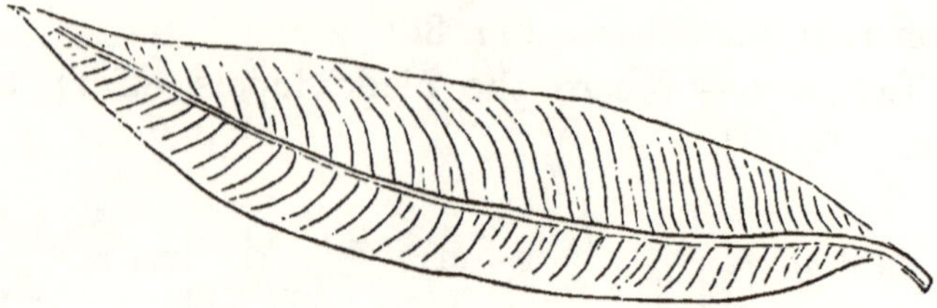

SCHEDULE THIRD, DESCRIBING FIG. 35.

Parts?	Blade, Petiole.
Venation?	Feather-veined.
Margin?	Entire.
Base?	Slightly tapering.

The base of Fig. 35 is much less tapering than Fig. 31. You will find all degrees, in this respect, from very blunt to very tapering. You will also be likely to find many leaves to which none of these pictures apply. In such cases you may write, I do not know, in the schedule, and wait till further exercises have shown you how to describe them.

---

sources shall be as extensive as possible. Teachers should therefore press beginners and negligent pupils about looking for new specimens. After a little time, such pressing will, in most cases, be unnecessary; for, when the interest and pride of a child are awakened by success in describing plants, he will take increasing pains to find new subjects for description.

EXERCISE VIII.

Apexes.

THE APEX of a leaf is its top, or free end.

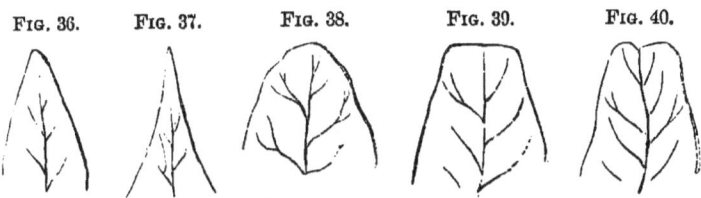

FIG. 36.   FIG. 37.   FIG. 38.   FIG. 39.   FIG. 40.

The Apex of a leaf may be:

ACUTE'.—Simply ending with a point. Fig. 36.

ACU'MINATE.—Ending with a long tapering point. Fig. 37.

OBTUSE'.—Blunt. Fig. 38.

TRUN'CATE.—Cut off at the apex. Fig. 39.

RETUSE'.—With the end rounded inward. Fig. 40.

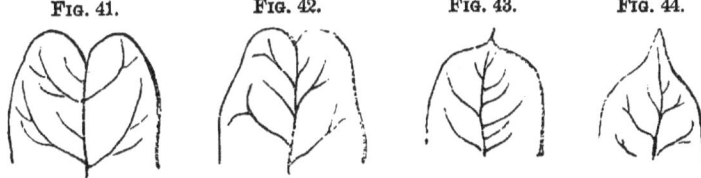

FIG. 41.   FIG. 42.   FIG. 43.   FIG. 44.

OBCOR'DATE.—Heart-shaped at the apex. Fig. 41.

EMAR'GINATE.—With a small notch at the apex. Fig. 42.

MU'CRONATE.—Tipped with a stiff, sharp point. Fig. 43.

CUS'PIDATE.—Suddenly ending with a sharp, slender point. Fig. 44.

The words acute, acuminate, and obtuse may be used to describe *bases* as well as *apexes*, and, when we wish to say that a shape is less acute or less acuminate than Figs. 36 and 37, we may say it is sub-acute or sub-acuminate, as in the schedule to Fig. 45.

Fig. 45.

Schedule Fourth, describing Fig. 45.

Parts?	Blade, Petiole, Stipules.
Venation?	Net-veined, Feather-veined.
Margin?	Serrate.
Base?	Obtuse.
Apex?	Sub-acute.

THE LEAF.

EXERCISE IX.

Forms of Lobes.

THE most striking difference in lobed leaves is the one seen in contrasting Fig. 46 with Fig. 47. It will be quite enough to ask of young beginners that they report whether the lobes of a leaf are rounded or acute.

FIG. 46.   FIG. 47.

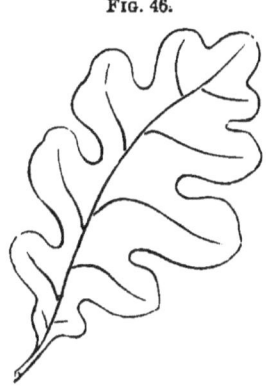

Rounded Lobes.   Acute Lobes.

But there may be older pupils who could profitably go further in observing the lobes of leaves. They vary much in size and shape, and are rarely all alike upon the same leaf. The lobe at the apex of a leaf is called the *Terminal* lobe, and is usually unlike all the others. The two lobes at the base are called *basal* lobes, and these also are usually unlike all the rest; for any pupils who would desire fuller observations upon lobes, a schedule might be prepared with two additional lines and the two questions, Terminal? and Basal? added to the present one. It might be well in such a case to give the number of lobes upon the leaf, along with their form, in answer to the question Lobes? While the peculiarities of the terminal and basal lobes would be given after these questions.

FIG. 48.

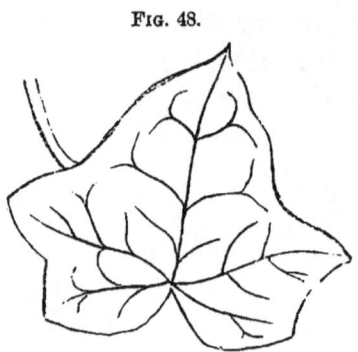

SCHEDULE FIFTH, DESCRIBING FIG. 48.

Parts?	Blade, Petiole.
Venation?	Palmate-veined.
Margin?	Entire.
Base?	Cordate.
Apex?	Acute.
Lobes?	Acute and Sub-acute.

In dealing with lobed leaves, you will not always find the base and apex so easily described as is Fig. 53, in the schedule. If they give you trouble, you may omit the questions, Base? and Apex?

THE LEAF.

EXERCISE X.

Forms of Sinuses.

The Si'nus of a leaf is the space left between lobes.
We represent here some of the most usual forms presented by Sinuses, with the terms describing them printed below the pictures.

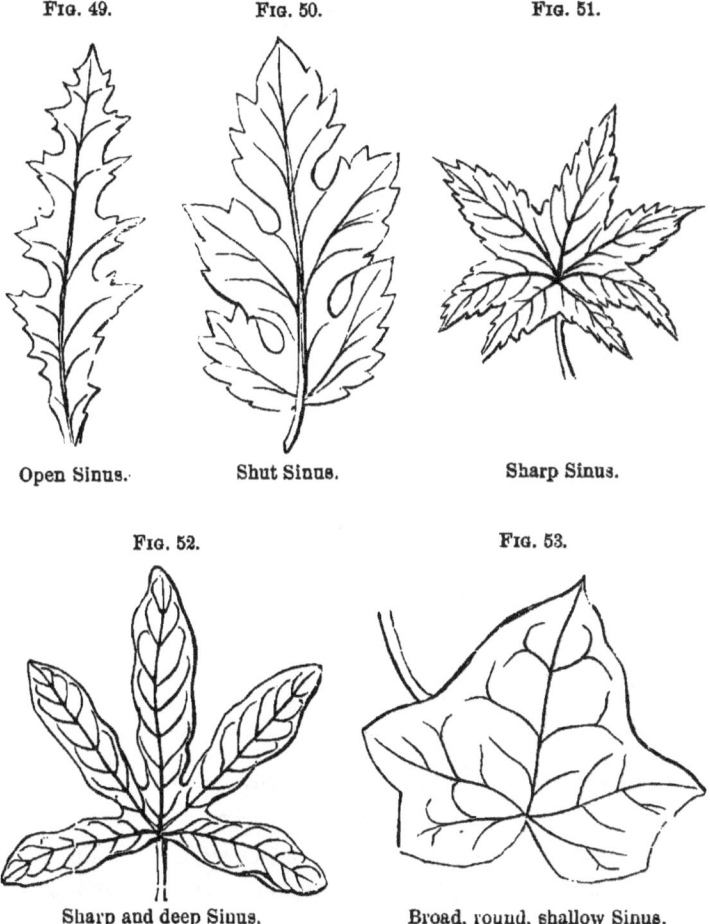

Fig. 49. Open Sinus.

Fig. 50. Shut Sinus.

Fig. 51. Sharp Sinus.

Fig. 52. Sharp and deep Sinus.

Fig. 53. Broad, round, shallow Sinus.

THE FIRST BOOK OF BOTANY.

Fig. 54. Fig. 55.

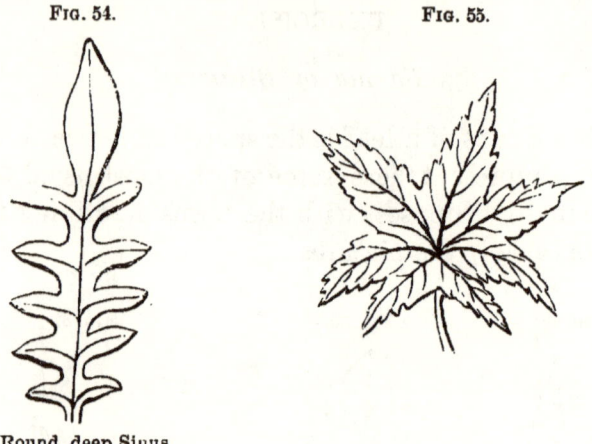

Round, deep Sinus.

Schedule Six, describing Fig. 55.

Parts ?	Blade, Petiole.
Venation ?	Palmate-veined.
Margin ?	Serrate.
Base ?	A broad, open Sinus.
Apex ?	Acute.
Lobes ?	Acute.
Sinuses ?	Sharp, upper ones deep.

Note.—It will be observed that our exercises contain none of the descriptions of plants and explanations of their growth which usually make up the text of botanies. These might be

THE LEAF.

EXERCISE XI.

Kinds of Leaves.

FIG. 56.   FIG. 57.   FIG. 58.

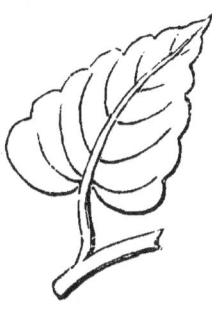

Sessile Leaf.   Stipulate and Petiolate Leaf.   A Petiolate and Exstipulate Leaf.

A Ses′sile Leaf is a leaf without a petiole.
A Stip′ulate Leaf is a leaf that has stipules.
A Pet′iolate Leaf is a leaf that has a petiole.
An Exstip′ulate Leaf is a leaf without stipules.

In Schedule Seven, it will be seen, we have dropped the question Parts? and put Kind? in its place. The words by which you answer this question are very long, but you can soon learn to handle them, and by-and-by you will find them much more convenient in leaf-description than it will be always to give a list of the parts.

---

easily given, but it would be a departure from our essential plan. The work before us—the observation of the external characters of plants—is itself extensive, and it can only be well done by making it at first our sole occupation. To observe carefully, to repeat our observations till they are familiar, and to acquire the ready and accurate use of the vocabulary of description, are the only true foundation of a knowledge of botany;

Fig. 59.

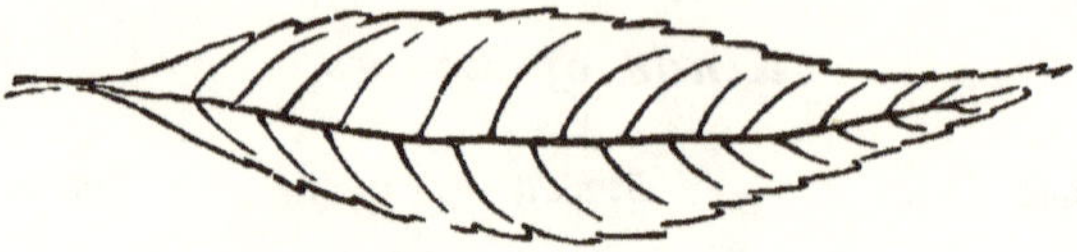

Schedule Seven, describing Fig. 59.

Kind ?	*Petiolate, Exstipulate.*
Venation ?	*Feather-veined.*
Margin ?	*Serrate.*
Base ?	*Acute.*
Apex ?	*Acuminate.*
Lobes ?	
Sinuses ?	

and we must be careful not to anticipate the work which belongs to a higher stage of the pupil's progress. The accounts of tissues, structures, and functions, add nothing to the understanding of plant-forms, and they afford proper subjects for future exercises in observation, to be given in a second book. What we have presented is eminently adapted to childhood, when sense-impressibility, and curiosity about appearances are strongest, and before the reflective powers are much developed.

The apparent meagreness of these pages is, therefore, intentional. They might easily have been filled with interesting reading matter about plants, but that would have opened the door to lesson-learning and reciting, which is a thing we specially wish to prevent.

THE LEAF.

EXERCISE XII.

Shapes of Leaves.

COMPARE leaves that are not lobed with the first three groups of pictures.

LEAVES THAT ARE BROADEST IN THE MIDDLE.

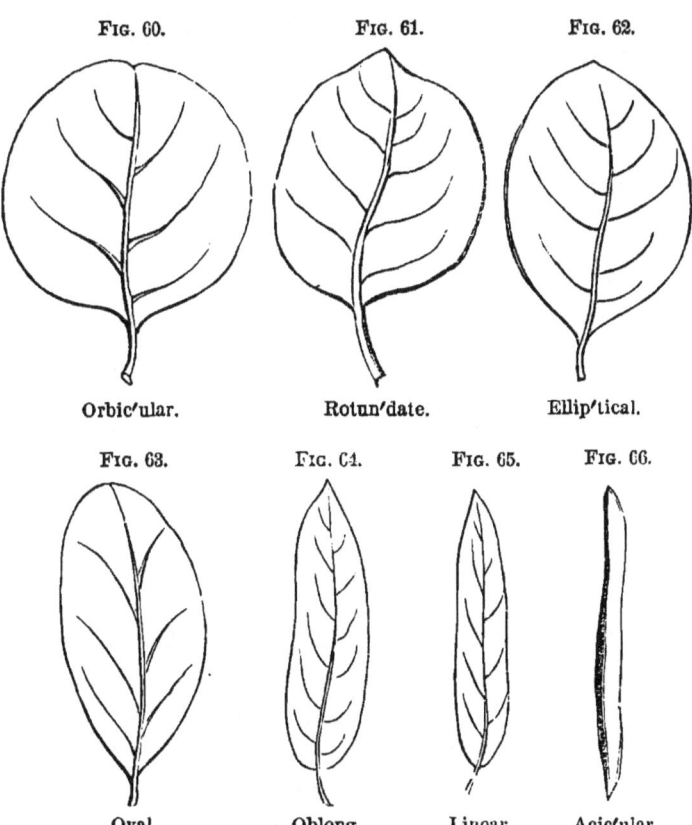

Fig. 60. Orbic′ular.
Fig. 61. Rotun′date.
Fig. 62. Ellip′tical.
Fig. 63. Oval.
Fig. 64. Oblong.
Fig. 65. Linear.
Fig. 66. Acic′ular.

LEAVES THAT ARE BROADEST AT BASE.

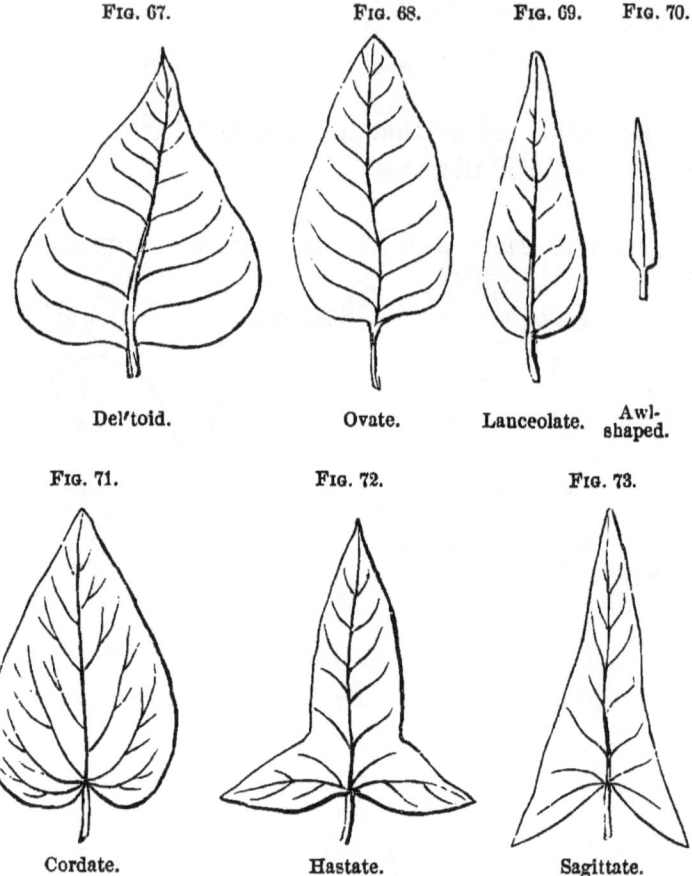

Fig. 67. Del'toid. Fig. 68. Ovate. Fig. 69. Lanceolate. Fig. 70. Awl-shaped.
Fig. 71. Cordate. Fig. 72. Hastate. Fig. 73. Sagittate.

Some of the names here applied to the whole leaf have already been used to describe a part of a leaf. For instance, among bases we had the heart-shaped base, and now a particular leaf-form is said to be heart-shaped. But it will soon be seen that heart-shaped bases may occur in leaves of very various forms, though there is one general form in which the entire leaf resembles a heart, and is therefore said to be heart-shaped. So the base of a leaf may look like an arrow, while the rest of it is very unlike an arrow. The apex may be truncate or obcordate, or any other form rather than the acute ending of an arrowshaped leaf. Follow the order of the schedule carefully in your descriptions till you begin to grow familiar with varying leaf-forms, and soon all appearance of confusion in the use of words will be at an end.

LEAVES THAT ARE BROADEST AT THE APEX.

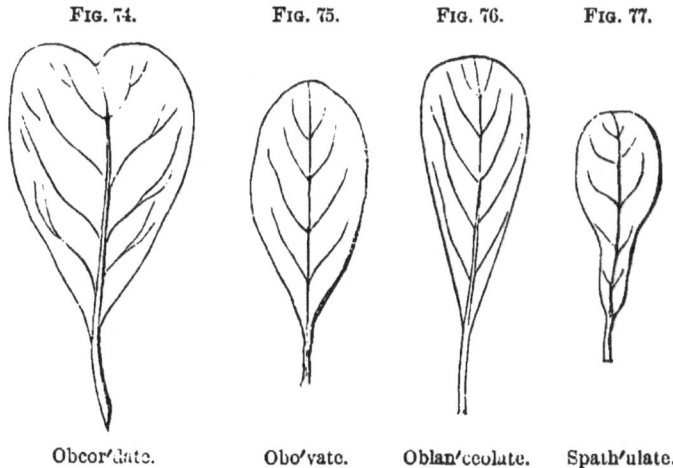

FIG. 74. FIG. 75. FIG. 76. FIG. 77.

Obcor′date. Obo′vate. Oblan′ceolate. Spath′ulate.

Do not expect to find an exact reproduction in Nature of the forms pictured in the book. You are simply to see which of the pictures your leaf is nearest like, and give it the name or the combination of names which the comparison seems to justify.

There are, of course, many leaves that you will not at first be able to describe. But if you find a very puzzling leaf, to which the schedule does not seem to apply, you may compare it with the following pictures. Perhaps it will be like one of these, and if so, if you cannot describe it, you can at least learn what to call it. If it is not like any of these pictures, it will be best to postpone its study for the present. By-and-by you will know better how to manage it.

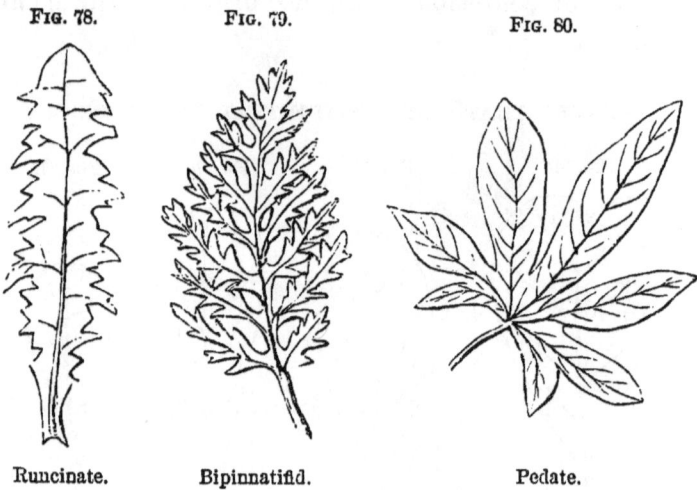

Fig. 78.   Fig. 79.   Fig. 80.

Runcinate.   Bipinnatifid.   Pedate.

A Run'cinate leaf is a lobed feather-veined leaf, in which the lobes point backward toward the base. Fig. 78.

Bipinnat'ifid leaves are formed when a deeply-lobed feather-veined leaf has its lobes again lobed, as in Fig. 79.

A Ped'ate leaf is a lobed palmate-veined leaf, in which the lobes at the base are lobed again, and give the leaf a look like the foot of a bird. Fig. 80.

Curled leaves (Fig. 81) are formed by a spreading of the border of the blade.

THE LEAF. 43

FIG. 81.

FIG. 82.   FIG. 83.   FIG. 84.

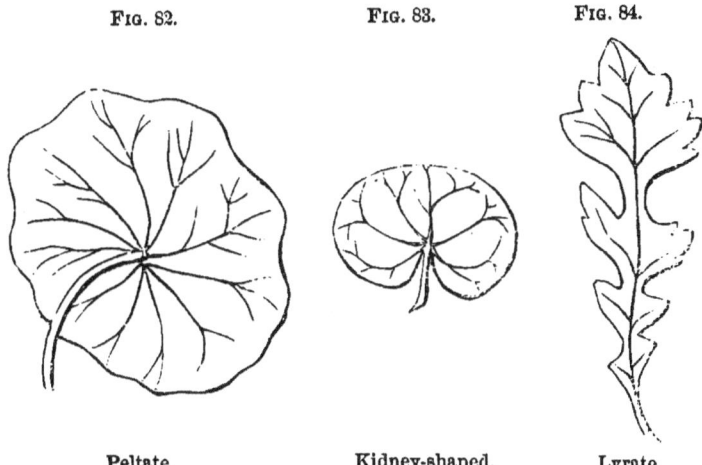

Peltate.   Kidney-shaped.   Lyrate.

PELTATE leaves are round, and have the petiole attached near the middle of the under surface of the blade. Fig. 82.

A KIDNEY-SHAPED leaf is short and broad, with a rounded apex and heart-shaped base. Fig. 83.

A LYRATE leaf is a lobed feather-veined leaf, with the terminal lobe much larger than the others. Fig. 84.

LACINIATE leaves are so named because they look as if they had been gashed with scissors. Fig. 85 is an example of such a leaf.

FIG. 85.

SCHEDULE SEVEN, DESCRIBING FIG. 85.

Kind?	*Petiolate, Exstipulate.*
Venation?	*Palmate-veined.*
Margin?	*Entire.*
Base?	
Apex?	
Lobes?	*Acuminate.*
Sinuses?	*Sharp, deep.*
Shape?	*Laciniate.*

EXERCISE XIII.

Petioles, Surfaces, and Colors.

The following schedule has three new questions added to it. The first is Petiole? The shape of the petiole, whether round, roundish, or half-round, should be observed, and written down. And if it be remarkable for its length or shortness, if it be unusually limber, or unusually stiff, you must mention these peculiarities about it.

COLOR? To this question the answer is easy. Leaves are sometimes light green, sometimes dark green; and sometimes the upper surface is one color, and the lower another. There are spotted and striped leaves, and some leaves have a brownish or reddish tinge. All these things are to be noted when you see them.

SURFACE? Observe whether the surface of a leaf has hairs or not. If it has hairs, write *hairy* after this question; but, if it has no hairs, write *glabrous*, which means free from hairs.

Again, surfaces are either *smooth* or *rough*, observe which, and write the result in the schedule.

Some leaves have a very *shiny* surface, and some are very *dull*, and these differences should be observed, and written down; but these qualities need not be noted unless they are strongly marked.

These characters cannot be conveniently represented by pictures, but they are readily seen in actual leaves. Feeling sure that you can easily make them out, we have not attempted to describe a leaf in schedule eight.

The schedule is now made up of the following questions :

SCHEDULE EIGHT.

Kind ?	
Venation ?	
Margin ?	
Base ?	
Apex ?	
Lobes ?	
Sinuses ?	
Shape ?	
Petiole ?	
Color ?	
Surface ?	

NOTE.—While in a book we must present one definite order of exercises, it is well if teachers use their own judgment in adhering to this order. Often, doubtless, much will be gained by judicious deviation. There are minds that demand variety, or their interest flags; and the minds of children, especially, are liable to grow weary of continued attention to one class of objects. Before proceeding with the exercises upon compound leaves, it may, therefore, be advisable to turn to the chapter upon the Inflorescence, or that upon the Flower, and occupy a little time with the opening exercise in which the names of parts are brought before the mind. The identification and naming of the parts of the flower will be easier to most children than the discrimination of simple and compound leaves; while dealing with another and more showy portion of the plant will stimulate the attention.

The use of schedule eight should, however, still be kept up, and, after a little while, the pupil will come back to the study of compound leaves with a fresh relish for the subject.

EXERCISE XIV.

Simple and Compound Leaves.

Fig. 88.   Fig. 89.

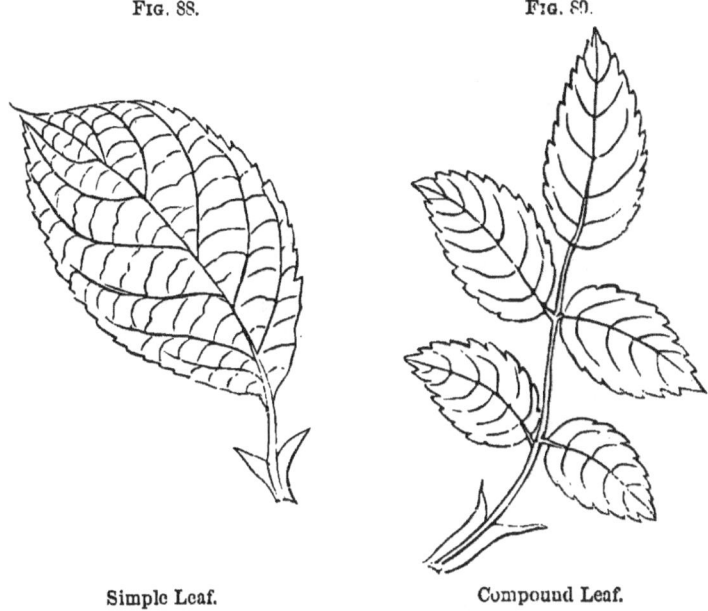

Simple Leaf.   Compound Leaf.

SIMPLE LEAVES have only one blade.

COMPOUND LEAVES consist of several distinct blades, called leaflets. You may know leaflets from lobes by their being entirely separate from each other.

UP to this time I suppose that pupils have described leaflets as leaves; but they must now be careful not to make this mistake. Let them confine themselves to simple leaves in using schedule eight, and write *simple leaf* upon it, to show that they have considered the matter.

It is sometimes a very nice point to decide between a deeply-lobed leaf and a compound leaf. If confusion at first arises, it must be patiently borne. We might add to the above definition of a compound leaf, that leaflets are jointed to the stalk, while the divisions called lobes never are. Such a statement would save trouble at first, but it would make greater trouble in the end. The truth is, that deeply-lobed leaves pass by insensible gradations into compound leaves, and compound leaves have their leaflets in all stages of connection with the common stalk, from a complete continuation of one into the other, up to a perfectly-jointed connection.

If the green matter of a leaf is continuous around the veins and along the ribs, however narrow the strip may be, it is quite correct to call such a specimen a simple leaf.

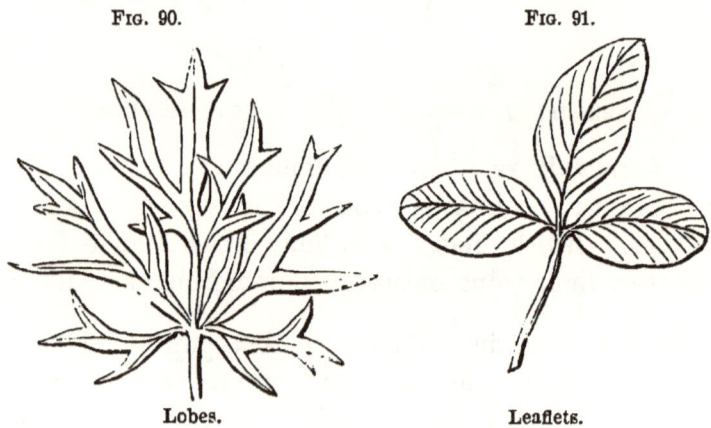

Fig. 90. Lobes.   Fig. 91. Leaflets.

There is no way, for the pupil, out of this difficulty except through a course of careful observation. Doubtless many mistakes will be made ; but mistakes are very useful in education.

COMPOUND LEAVES. 49

EXERCISE XV.

Parts of Compound Leaves.

Fig. 92.

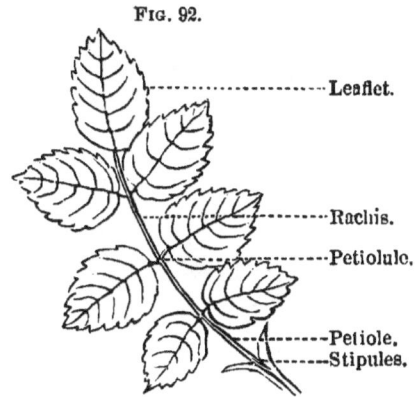

Fig. 93.

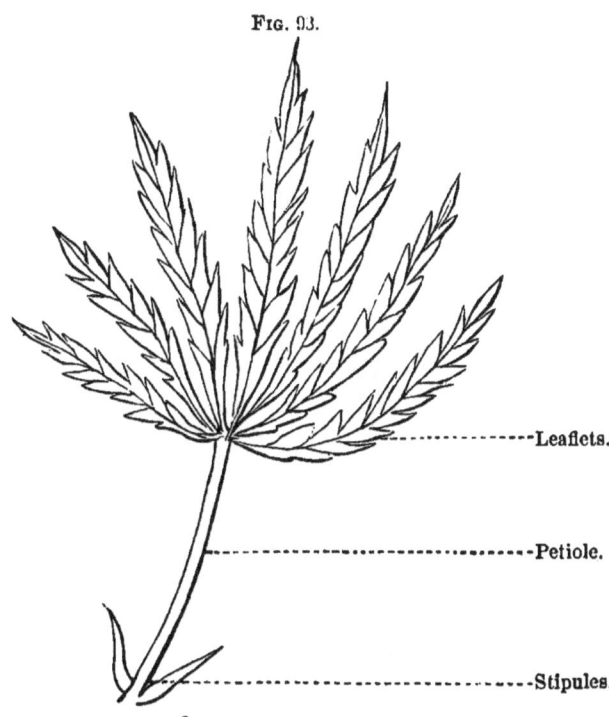

Fig. 94.

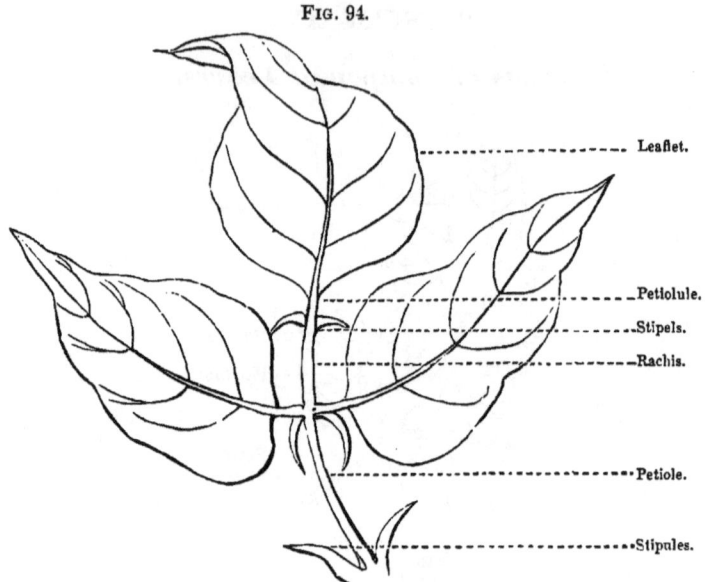

LEAFLET.—One of the blades of a compound leaf.
PET'IOLULE.—The stem of a leaflet.
STI'PELS.—The stipules of leaflets.
RA'CHIS.—The continuation of the petiole to which leaflets are attached.

Fig. 95.

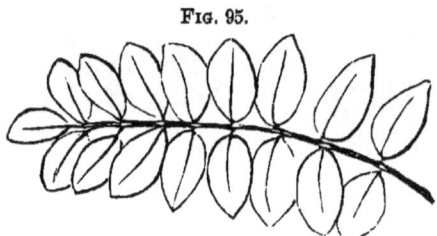

SCHEDULE NINE, DESCRIBING FIG. 95.

Parts.	Rachis, Petiole, Leaflets.
No. of Leaflets.	17.

COMPOUND LEAVES.

EXERCISE XVI.

Pinnate and Digitate Leaves.

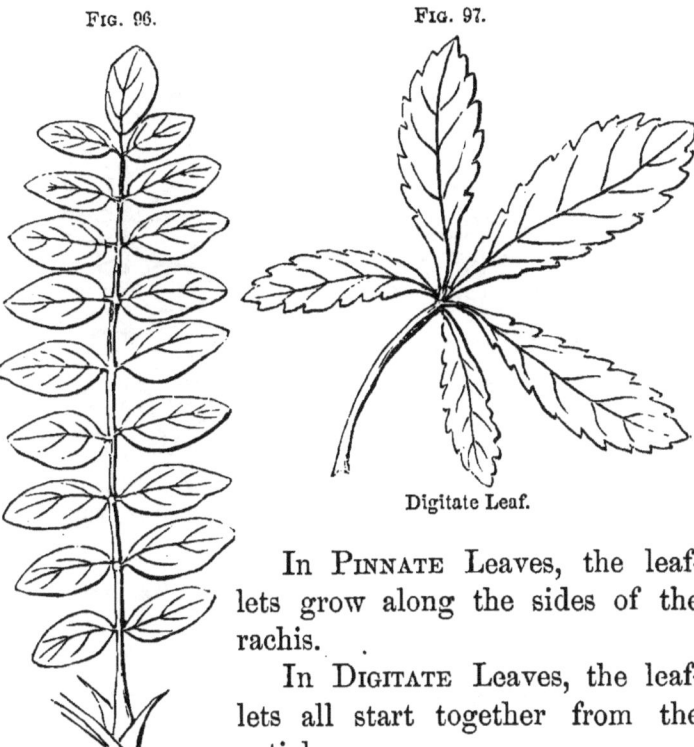

Fig. 96. Pinnate Leaf.
Fig. 97. Digitate Leaf.

In PINNATE Leaves, the leaflets grow along the sides of the rachis.

In DIGITATE Leaves, the leaflets all start together from the petiole.

SCHEDULE TEN, DESCRIBING FIG. 97.

Parts ?	*Petiole Leaflets.*
No. Leaflets ?	*5.*
Kind ?	*Digitate.*

EXERCISE XVII.

Varieties of Pinnate Leaves.

Fig. 98.

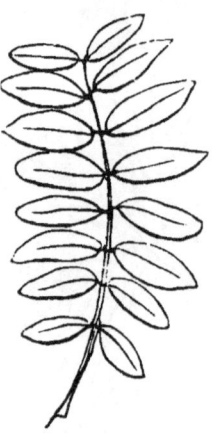

Abruptly Pinnate.

Fig. 99.

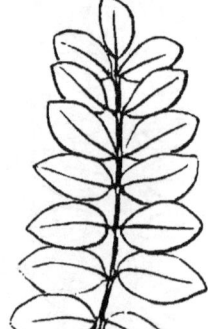

Unequally Pinnate.

Fig. 100.

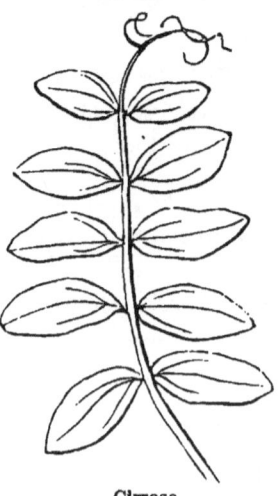

Cirrose.

Fig. 101.

Interruptedly Pinnate.

COMPOUND LEAVES. 53

ABRUPTLY PINNATE.—When the leaf terminates in a *pair* of leaflets. Fig. 98.

UNEQUALLY PINNATE.—When the leaf terminates in an *odd*, or single, leaflet. Fig. 99.

CIRROSE.—When the rachis ends in slender branching curls, called *tendrils*. Fig. 100.

INTERRUPTEDLY PINNATE.—When the leaflets are alternately large and small. Fig. 101.

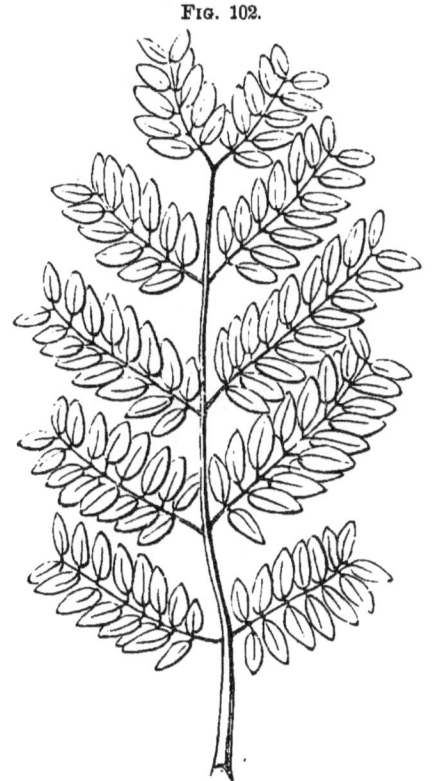

FIG. 102.

Twice Pinnate.

TWICE PINNATE.—When the petiolule is continued as a rachis which bears the leaflets.

FIG. 103.

Thrice Pinnate.

THRICE PINNATE.—When the leaflets are borne upon a third rachis, branching off from the second.

EXERCISE XVIII.

Varieties of Digitate Leaves.

THREE-FINGERED.—A digitate leaf with three leaflets.

COMPOUND LEAVES.

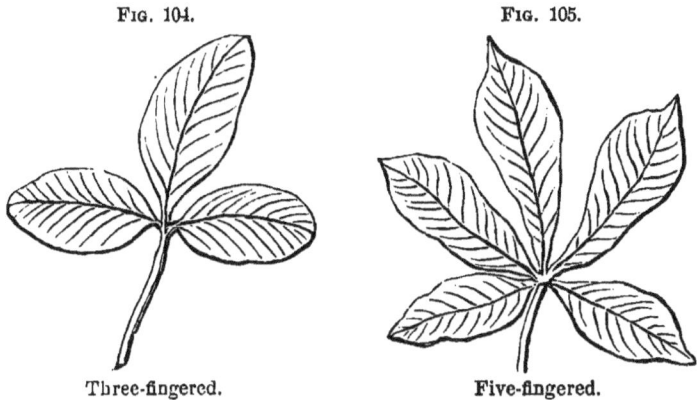

Fig. 104. Three-fingered.

Fig. 105. Five-fingered.

FIVE-FINGERED.—A digitate leaf with five fingers or leaflets.

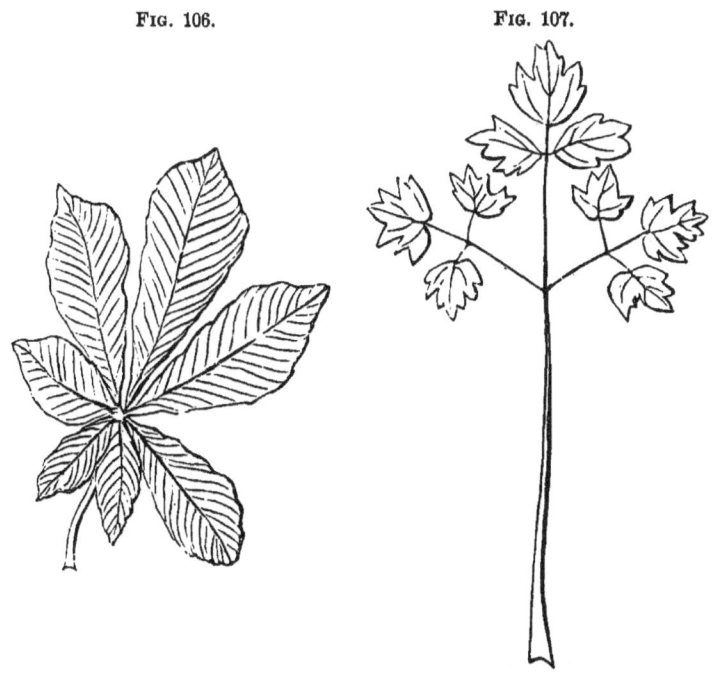

Fig. 106. Seven-fingered.

Fig. 107. Twice Three-fingered.

56   THE FIRST BOOK OF BOTANY.

SEVEN - FINGERED.—A digitate leaf with seven fingers or leaflets. Fig. 106.

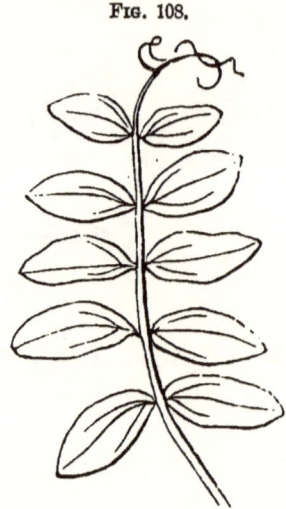

FIG. 108.

SCHEDULE ELEVEN, DESCRIBING FIG. 108.

Parts?	Petiole, Rachis, Leaflets.
No. Leaflets?	10.
Kind?	Pinnate.
Variety?	Cirrose.

NOTE.—This is the last leaf-schedule. With the next chapter we begin the study of the stem. But we must still in some way pursue the study of leaf-forms, if we would render permanent the knowledge we have already acquired. An observation is by no means a mental possession as soon as it is made. True knowledge is always a growth requiring *time;* and observations have not only to be *made,* but to be *repeated,* and

EXERCISE XIX.

Forms of Stipules.

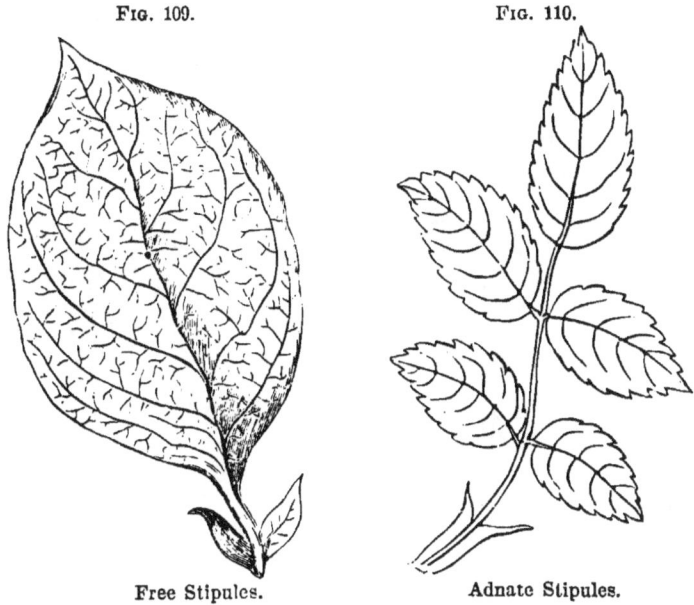

Fig. 109. Free Stipules.

Fig. 110. Adnate Stipules.

STIPULES are:

FREE.—When not united with any other part.

ADNATE.—When they grow to the petiole.

---

the facts knit into their places, to make them reliable mental possessions. Understanding a thing is but the first step toward its real acquirement. A succession of frequent observations is necessary to induce familiarity with objects, and there must also be a recurrence to them—a revival of impressions after considerable intervals of time. It is possible to have an *intense* familiarity with things observed, by occupying the whole consciousness with them for a short time, but effects thus produced are not lasting. We shall, therefore, continue our observations of leaves, and record them upon the stem-schedule. Pupils who have been diligent in the use of the

Fig. 111. Fig. 112.

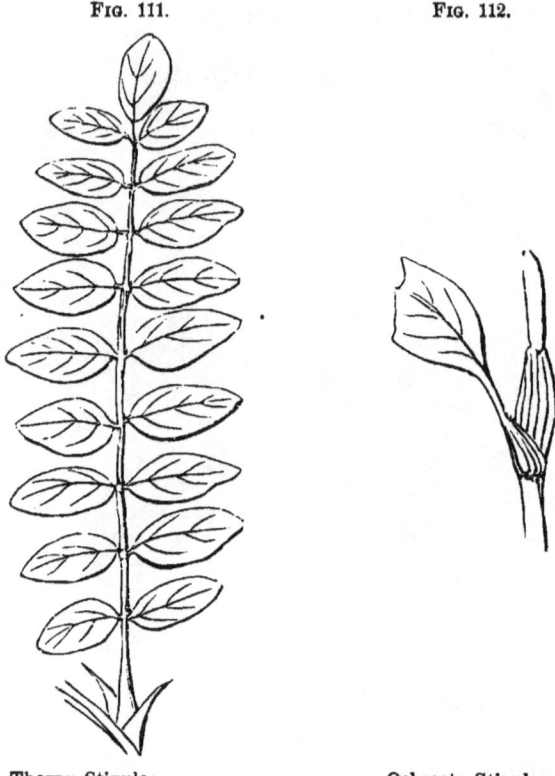

Thorny Stipules.   Ochreate Stipules.

THORNY.—Like thorns.

OCHREATE.—When they form a sheath around the stem.

If any of the distinctions among compound leaves bother very young pupils, let the observation of such be omitted for the present.

---

preceding schedules, ought now to be able to describe leaves without their aid. Exercise XX. consists of two descriptions, in which the schedule questions are omitted. The order of description which has all along been followed will naturally have become the order of thought with pupils, and the prompt-

COMPOUND LEAVES. 59

EXERCISE XX.

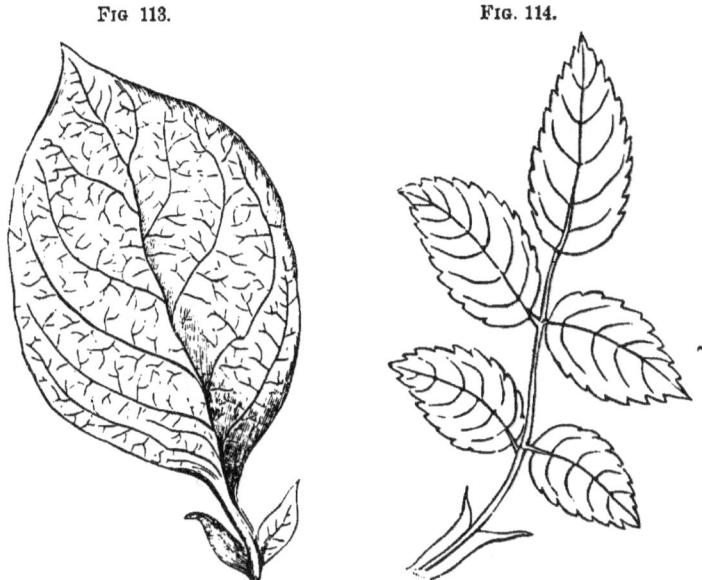

Fig 113.   Fig. 114.

LEAF (Fig. 113).—Simple, petiolate, stipulate, net-veined, feather-veined, entire, abruptly acuminate, broadly oval; *petiole*, short, bordered by the blade; *stipules*, free.

LEAF (Fig. 114).—Compound, petiolate, stipulate, unequally pinnate, number of leaflets, 5; *leaflets*, petiolate, feather-veined, serrate, ovate; *stipules*, adnate.

---

ing of questions is now unnecessary. They have answered their purpose if they have led to a knowledge of the parts of leaves and their most important modifications of form. When this is done, it will be much more important that the pupil be unassisted in making descriptions than that he be always methodical and correct.

For pupils that are old enough to punctuate their descriptions, the following rule will be useful:—1. Separate adjectives relating to the same noun, by commas; 2. Parts of the same organ, by semicolons; 3. Distinct organs, by a period.

CHAPTER II.

THE STEM.

EXERCISE XXI.

Parts of the Stem, and Leaf Axil.

FIG. 115.

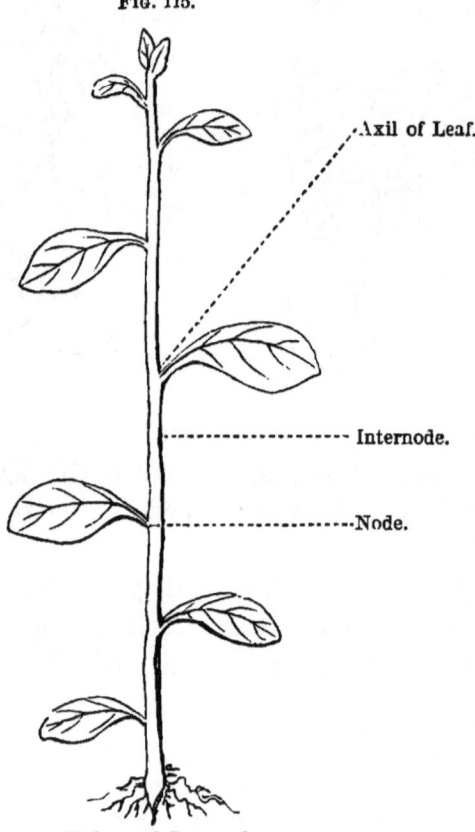

Nodes and Internodes.

THE STEM.

Fig. 116.

Fig. 117.

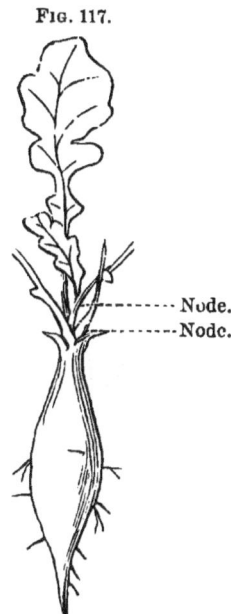

NODE.—The point on the stem from which leaves are given off.

INTERNODE.—The portion of the stem between two nodes.

LEAF AXIL.—The point at the upper side of the leaf where it joins the stem.

NOTE.—Children will easily find the nodes and internodes of most stems, but they should not, therefore, hurry past this exercise without tracing the successive internodes of many stems from the root upward. The teacher should also see that a clear idea is gained of the axil of a leaf.

If Figs. 116 and 117 are not intelligible to beginners, and the parts of short stems like these are distinguished with difficulty, let them be passed over, as the coming exercises are not dependent upon these discriminations. But, for those who can make them out, they will be profitable.

EXERCISE XXII.

Appendages of the Stem.

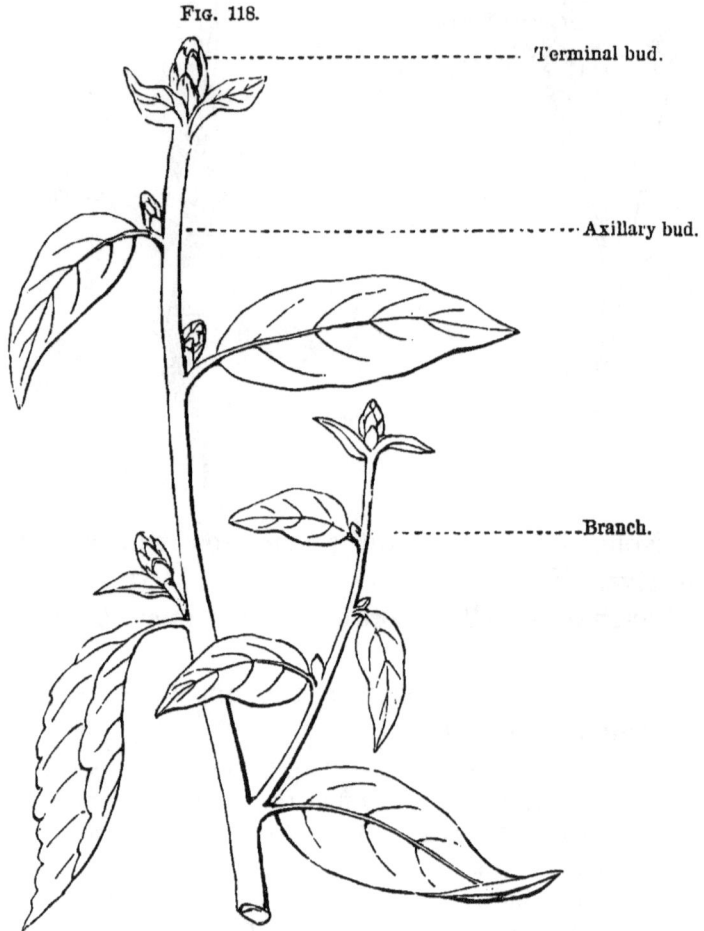

Fig. 118.

Terminal Bud.—The bud at the end of the stem.
Ax'illary Bud.—The bud in the axil of a leaf.
Branch.—A stem which starts from the axil of a leaf.

THE STEM.

Fig. 119.

Schedule Twelve, describing Fig. 119.

Parts?	Nodes, Internodes.
Appendages?	Leaves, Tendrils, Flowers.

Leaf.—Petiolate, exstipulate, palmate-veined, 5-lobed, broad as long; lobes rounded, entire; sinuses deep, round; *petiole* long, slender.

The appendages of the stem (Fig. 118) are leaves, buds, and branches. The terminal bud continues the growth of the main stem. Axillary buds give rise to branches, or secondary stems.

EXERCISE XXIII.

Position of Leaves.

Fig. 120. Cauline Leaves.   Fig. 121. Radical Leaves.

Cau′line leaves grow along the *caulis*, or stem.

Radical leaves start close to the ground, or below its surface.

Note.—The term *radical* would seem to imply that the leaves spring from roots, which is not the case, as shown in Fig. 117.

THE STEM.

Fig. 122.

Schedule Thirteen, describing Fig. 122.

Appendages?	*Leaves, Flowers.*
Leaf-position?	*Cauline.*

Leaf.—Simple, sessile, feather-veined, entire, lanceolate; *stipule*, ochreate.

The question, Parts? is now dropped, because it is answered in giving the position of the leaves. To say that leaves are *cauline* is to say that the stem is composed of both nodes and internodes, while, if the stem has radical leaves only, there are no internodes.

Note.—When the nodes of a stem are distinctly jointed, when they are swollen and watery (tumid), when they are hairy, or when of a different color from the internodes, they give a peculiar aspect to the plant, and pupils should be en-

66 THE FIRST BOOK OF BOTANY.

EXERCISE XXIV.

Arrangement of Leaves on the Stem.

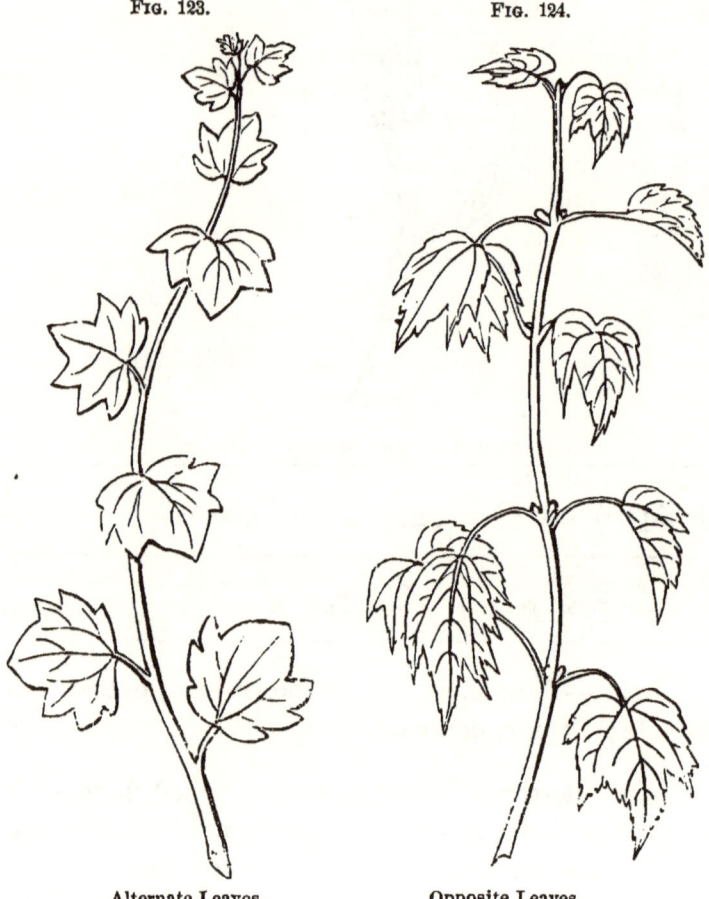

Fig. 123.

Fig. 124.

Alternate Leaves.   Opposite Leaves.

ALTERNATE LEAVES.—Leaves are alternate on the stem when there is but one at each node, as in Fig. 123.

---

couraged to record such facts upon the schedule. Very long or very short internodes, and other noticeable peculiarities, should

THE STEM. 67

OPPOSITE LEAVES.—When two leaves grow opposite each other, we call it the *opposite* arrangement. Fig. 124.

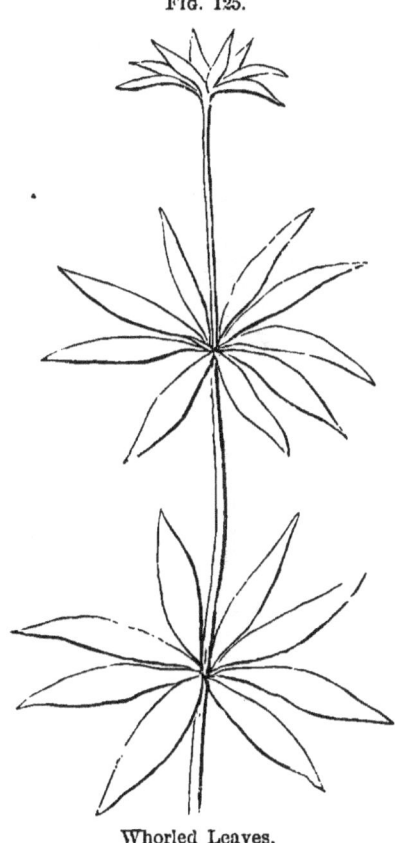

Fig. 125.

Whorled Leaves.

WHORLED LEAVES.—When there are more than two leaves at a node, we say the leaves are whorled.

---

be stated. A word or two at the bottom or back of the schedule, as, nodes tumid, or, internodes very long, is all that is requisite. Brevity and precision of statement should always be insisted upon.

68 THE FIRST BOOK OF BOTANY.

Fig. 126.

SCHEDULE FIFTEEN, DESCRIBING FIG. 126.

Appendages?	Leaves.
Leaf-position?	Cauline.
Leaf-arrangement?	Alternate.

THE LEAF.—Sessile, feather-veined, serrate, lanceolate.

NOTE.—Leaf-position and leaf-arrangement pertain as much to the leaf as to the stem, but observations concerning them could not be properly made until something was known of the stem. When the pupil becomes familiar with these characters, it will, perhaps, be more appropriate to notice them in the leaf-description than in the stem-description.

THE STEM.

EXERCISE XXV.

Shapes of Stems.

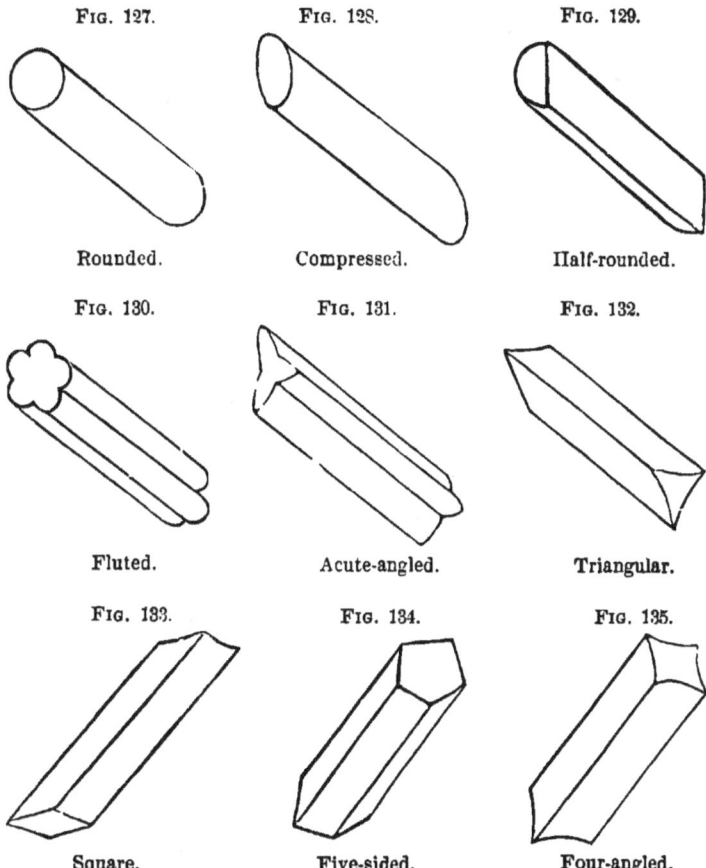

Fig. 127. Rounded. Fig. 128. Compressed. Fig. 129. Half-rounded.
Fig. 130. Fluted. Fig. 131. Acute-angled. Fig. 132. Triangular.
Fig. 133. Square. Fig. 134. Five-sided. Fig. 135. Four-angled.

These are by no means all the shapes, nor the precise shapes that stems assume, but their forms will most commonly be found to approach very nearly to some of these outlines. If any forms occur that are so widely different from the pictures as to perplex the pupil, he will consult the teacher.

70 THE FIRST BOOK OF BOTANY.

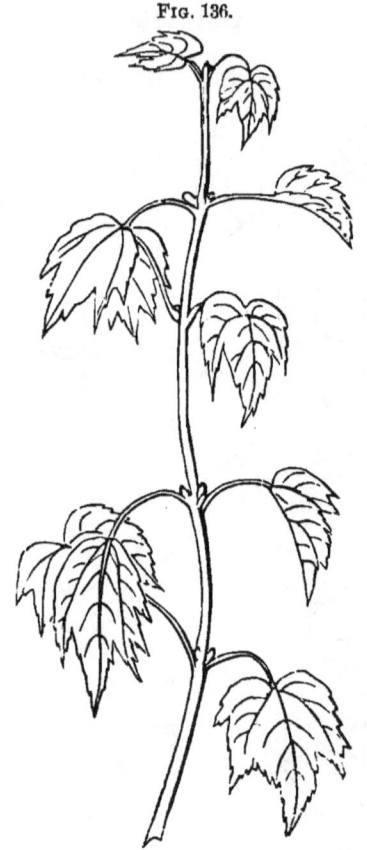

Fig. 136.

Schedule Sixteen, describing Fig. 136.

Appendages?	*Leaves, Buds.*
Leaf-position?	*Cauline.*
Leaf-arrangement?	*Opposite.*
Shape?	*Round.*

THE STEM. 71

LEAF.—Petiolate, exstipulate, palmate-veined, serrate, base cordate, 5-lobed, terminal lobe acuminate, leaf broader than long.

EXERCISE XXVI.

Attitude of Stems.

FIG. 137.   FIG. 138.

Erect.   Drooping.

ERECT stems stand upright.
DROOPING stems are limber, and bend over.

72   THE FIRST BOOK OF BOTANY.

Fig. 139.

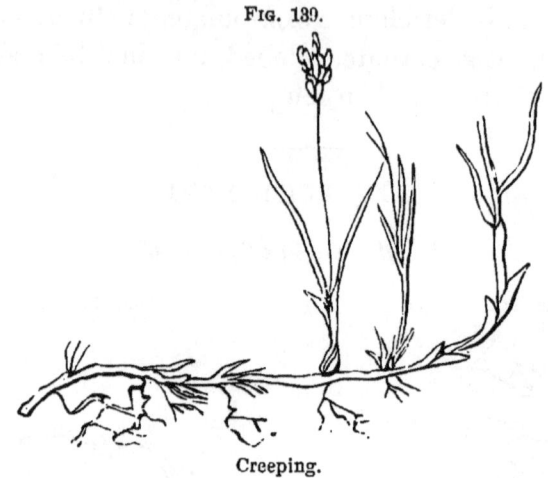

Creeping.

Fig. 140.

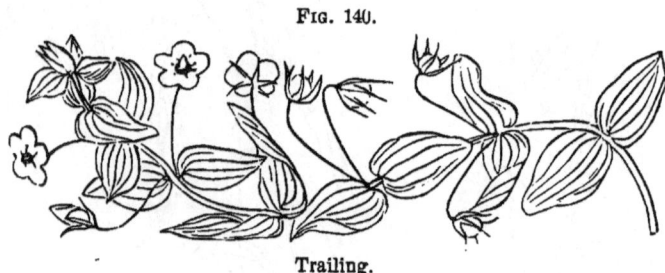

Trailing.

Fig. 141.

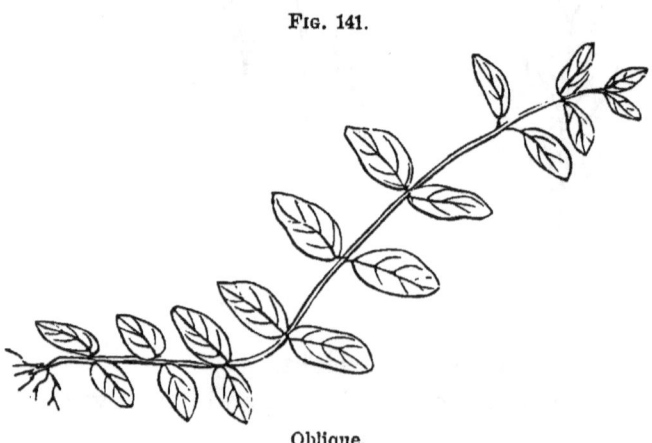

Oblique.

THE STEM. 73

CREEPING stems lie along the ground, and send down roots from their nodes. Fig. 139.

TRAILING stems are weak, and lie loosely along the ground. Fig. 140.

OBLIQUE stems stand slanting. Fig. 141.

FIG. 142.   FIG. 143.

Climbing.   Twining.

CLIMBING stems are weak, and cling by tendrils to the objects about them.

TWINING stems are too weak to stand alone, and support themselves by winding around other stems.

SCHEDULE SEVENTEEN, DESCRIBING FIG. 143.

Appendages?	Leaves, Flowers.
Leaf-position?	Cauline.
Leaf-arrangement?	Alternate.
Shape?	Round.
Attitude?	Twining.

LEAF.— Simple, petiolate, exstipulate, feather-veined, entire, cordate, sub-acuminate.

EXERCISE XXVII.

Color, Surface, Size, Structure.

COLOR.—Stems may be spotted, striped, green, brown, red, or purple.

SURFACE.—The surface of stems, like that of leaves, is smooth, rough, shiny, dull, hairy, and glabrous.

SIZE.—Stems may be high or low, slender or thick, and it is easy to determine these points.

STRUCTURE.—To find out the structure of a stem, you must break it, and observe first whether it is *hollow* or *solid*. Next examine it to ascertain if it have any tenacious threads; these are *woody* fibres, and, when present, they help to make the stem hard

THE STEM. 75

and tough. It is then called a WOODY stem. But, if it is soft and brittle, it is an HERBACEOUS stem. The stem schedule consists now of the following questions:

SCHEDULE EIGHTEEN.

Appendages?	
Leaf-position?	
Leaf-arrangement?	
Shape?	
Attitude?	
Color?	
Surface?	
Size?	
Structure?	

NOTE.—In schedule eighteen, as in schedule nine, no picture is described, because two of the questions now added, viz., Color? and Structure? relate to features that cannot be easily represented in a picture, while size and surface, as seen in nature, are so unlike pictorial presentations, that an example given here would be but a poor guide in schedule practise. The descriptive terms used in answering these questions are so familiar as not to need illustration.

CHAPTER III.

THE INFLORESCENCE.

INFLORESCENCE.—The way flowers are placed upon plants is called their *inflorescence*.

EXERCISE XXVIII.

Solitary and Clustered Inflorescence.

FIG. 144.

Solitary Inflorescence.

THE INFLORESCENCE. 77

FIG. 145.

Clustered Inflorescence.

SOLITARY INFLORESCENCE is where only one flower grows upon a flower-stem. Fig. 144.

CLUSTERED INFLORESCENCE is where several flowers grow from the same flower-stem.

Flowers, or flower-clusters without stems, are said to be *sessile*.

NOTE.—This and the following exercise should be dealt with in the same manner as the first exercises in the chapters upon the leaf and stem.

EXERCISE XXIX.

Parts of the Inflorescence.

Fig. 146.

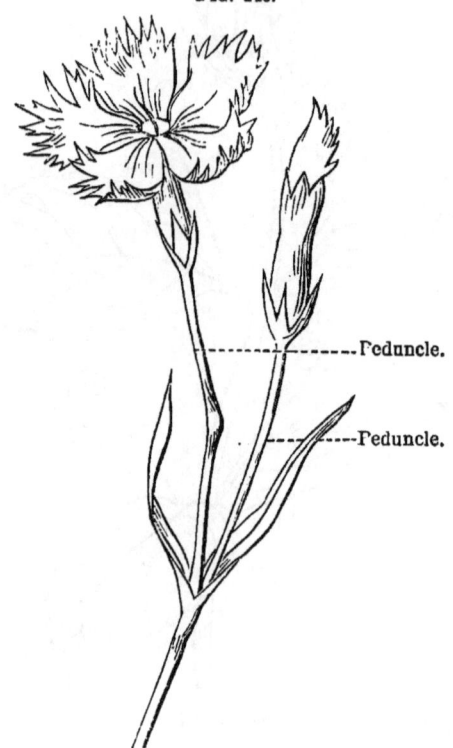

Pedun'cle.—The stem of a solitary flower, or of a flower cluster.

Bracts.—The small leaves of a flower-cluster on the peduncle, or rachis.

In'volucre.—A whorl of bracts.

Ped'icel.—One of the flower-stems in a cluster.

Bract'lets.—Very small leaves growing upon pedicels.

THE INFLORESCENCE. 79

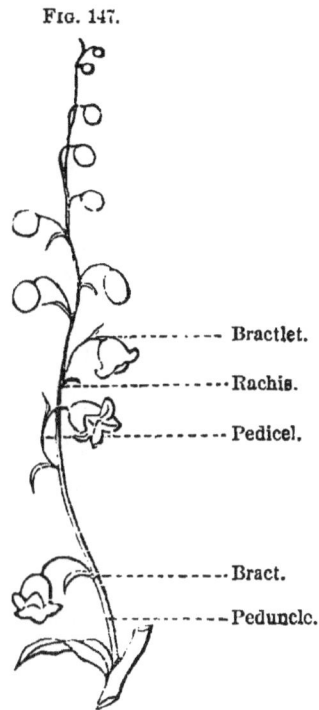

Fig. 147.

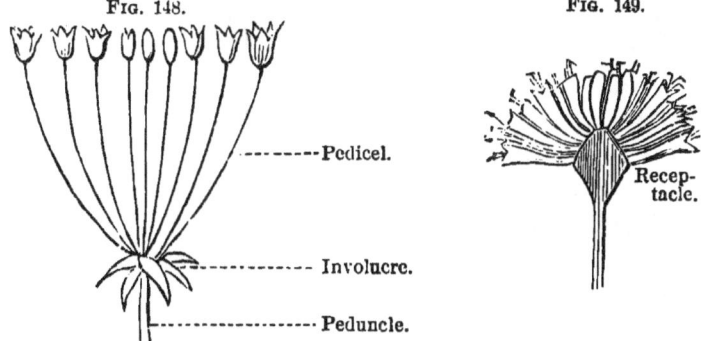

Fig. 148.   Fig. 149.

Ra′chis.—The continuation of a peduncle, from which flowers branch off.

Recep′tacle.—The top of a peduncle, from which several flowers start together.

EXERCISE XXX.

Attitude of Inflorescence.

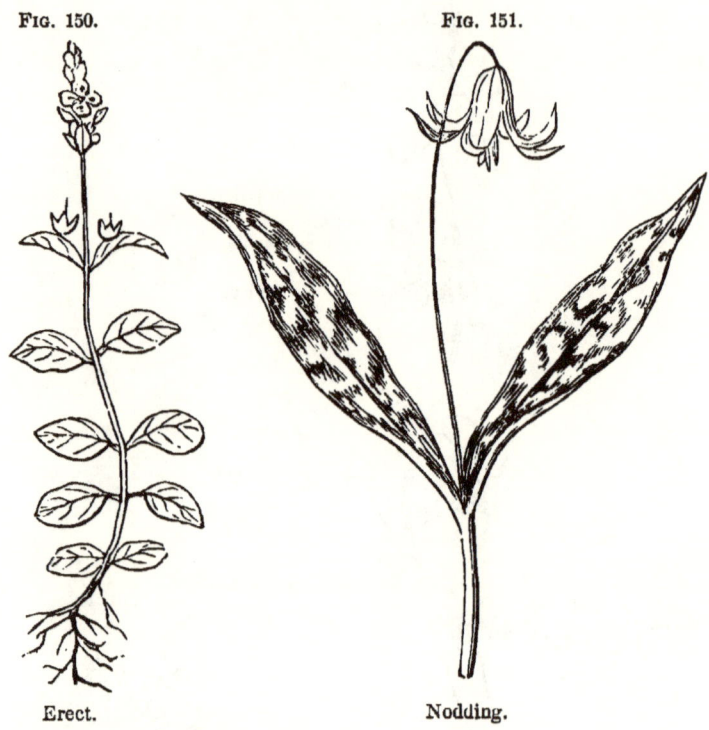

Fig. 150. Erect.
Fig. 151. Nodding.

ERECT.—Upright.
NODDING.—Bending over.

NOTE.—Many of the characters already noted as belonging to the stem of a plant, belong also to the peduncle. Its color, surface, shape, length, limpness, twist, and curvature, may be recorded in connection with the attitude in the same terms as are used in stem-descriptions.

Some of the statements in the description of Fig. 152 are to be compared with the living plant, and not the picture.

THE INFLORESCENCE. 81

Fig. 152.

Pendulous.

PEN'DULOUS.—Hanging down.

SCHEDULE EIGHTEEN, DESCRIBING FIG. 152.

Parts?	*Peduncle, Flower.*
Attitude?	*Pendulous.*

LEAF. — Simple, petiolate, exstipulate, feather-veined, irregularly-dentate, ovate-acuminate, green, smooth, cauline, opposite.

STEM.—Round, slightly bending, reddish brown, smooth, slender, solid, woody.

EXERCISE XXXI.

Solitary Terminal and Axial Inflorescence.

Fig. 153.

Solitary Terminal.

An inflorescence is SOLITARY TERMINAL when the stem, or branch, ends in a single flower.

The presence of nodes upon ordinary stems distinguishes them from flower-stems or peduncles.

THE INFLORESCENCE. 83

Fig. 154.

Solitary Axial.

A SOLITARY AXIAL flower is one where the peduncle starts from the axil of a leaf.

In Fig. 154 the peduncle of the lowest flower starts from the axil of the leaf, it is hence an axial flower; but the peduncle of the lowest flower in Fig. 153 starts at the first node of the branch, the growth of which it terminates; it is hence a terminal flower.

EXERCISE XXXII.

Clustered Axial and Terminal Inflorescence.

Fig. 156.

Clustered Terminal.

A TERMINAL CLUSTER of flowers is one that ends the growth of a stem, or branch. Fig. 156.

Observe that the lowest bud in Fig. 156 is hardly discernible as a flower-bud. The next is a little more advanced, the third still more, and so on till, at the top of the cluster, you see a fully expanded flower. The oldest flowers are at the top or centre of the cluster. This order is often reversed, the oldest flowers being at the bottom or outside of the cluster, and it is important for you to notice this circumstance in studying inflorescence.

THE INFLORESCENCE.

Fig. 157.

Clustered Axial.

An Axial Cluster of flowers is one where the peduncle starts from a leaf axil.

The question, Position? is now added to the inflorescence-schedule. Every inflorescence is either terminal or axial, and the pupil is to determine this point, in order to answer the new question. When he begins the study of botany in its higher aspects, he will find that much depends upon his having carefully observed such points as these.

SCHEDULE NINETEEN, DESCRIBING FIG. 157.

Parts?	*Peduncle, Pedicels, Flowers.*
Attitude?	*Erect.*
Position?	*Axial.*

LEAF.—Cauline, opposite, simple, sessile, feather-veined, crenate, or crenate-serrate, lower leaves, sub-acute, upper ones obtuse, lower leaves broadly ovate, upper ones broadly oval.

STEM round, erect, herbaceous.

NOTE.—Determination of the position of an inflorescence is often very easy, yet sometimes it is puzzling and difficult. For instance, although the cluster (Fig. 156) is clearly *terminal*, a thoughtful child might notice that each flower in this cluster is *axial*, and so hesitate in deciding how to describe it. Such perplexities will be gradually cleared up as the child advances with the study. It should be remembered that many of the observations begun with this book are necessarily incomplete. Cloudiness of perception concerning some matters must, therefore, be tolerated at first. Clear and complete ideas can only arise as the mind develops, and the subject is further pursued. There are portions of almost every study over which children are liable to get confused at first. They see difficulties, but cannot see through them. Yet the discovery of difficulties is as much a part of education as the discovery of facts. It is the overcoming of difficulties, and this mainly, that exercises the judgment, and calls forth mental power. But, to gain these ends, it is important that the child be left to himself. It is better for him to form his own opinion, even though it be wrong, than to have every thing explained in advance. Extended observation and continued thought may be trusted to correct errors made at first, as, without these conditions, there can be little real improvement.

THE INFLORESCENCE. 87

EXERCISE XXXIII.

Definite and Indefinite Inflorescence.

Fig. 158.

Definite.

ALL *solitary terminal* inflorescence, and all terminal clusters that, like Fig. 156, have their oldest flowers at the top or centre of the cluster, are said to be DEFINITE, because they end the growth of the stem or branch that bears them.

Fig. 159.

Indefinite.

All *axial* inflorescence is INDEFINITE, because the stem and branches, if there be any, may grow on just the same as before blossoming. The inflorescence in Fig. 159 is indefinite. The stem does not end with flowers, but with a leaf-bud, which continues its growth.

The question, Kind? is now added to the inflorescence-schedule, and pupils will state, in answer, whether the inflorescence is definite or indefinite.

THE INFLORESCENCE. 89

FIG. 160.

SCHEDULE TWENTY, DESCRIBING FIG. 160.

Parts?	Peduncle, Flowers, Rachis.
Attitude?	Erect.
Position?	Axial.
Kind?	Indefinite.

LEAF.—Cauline, opposite, simple, sessile, feather-veined, entire, oval.

STEM round, erect, slender, herbaceous.

NOTE.—Compare Fig. 160 with Fig. 156, and observe that they differ in the order in which the flowers unfold. In Fig. 160 the oldest flowers are the lowest in the cluster. There is no flower at the top of the cluster, ending the growth of the

EXERCISE XXXIV.

Varieties of Inflorescence.

SIMPLE.

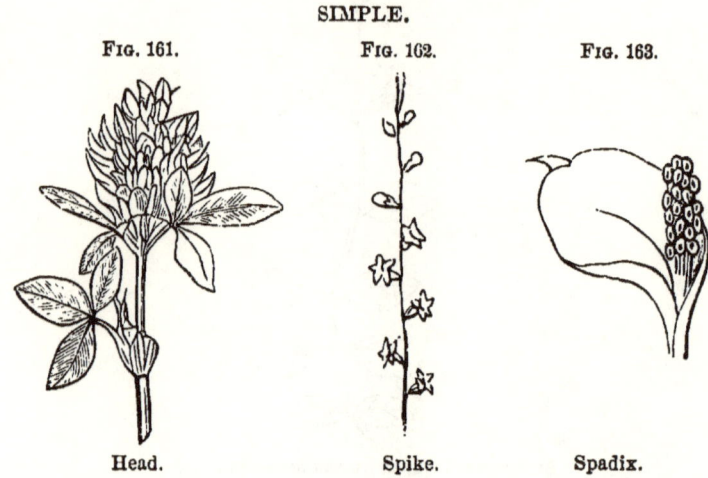

Fig. 161. Head. Fig. 162. Spike. Fig. 163. Spadix.

HEAD.—A more or less globular cluster of flowers, sessile upon the receptacle.

SPIKE.—A cluster of flowers, sessile upon a rachis.

SPA'DIX.—A spike with a thick rachis, and covered around by a single large leaf, or bract, called a spathe.

---

stem, and so, as the rachis may grow on, sending off flowers from its side, we say the inflorescence is indefinite.

The primary, or main stem, of a plant sometimes ends definitely, or with a flower, while the branches, or secondary stems, grow on, or are indefinite. Sometimes the main stem is indefinite, and the branches are definite. When both kinds of inflorescence are found upon the same plant, it should be stated.

To determine whether a flower-head is definite or indefinite, observe whether the unopened flowers are at the top or on the lower part of the cluster. It is only in rare instances that they all open so nearly at the same time as to show no differ-

THE INFLORESCENCE.

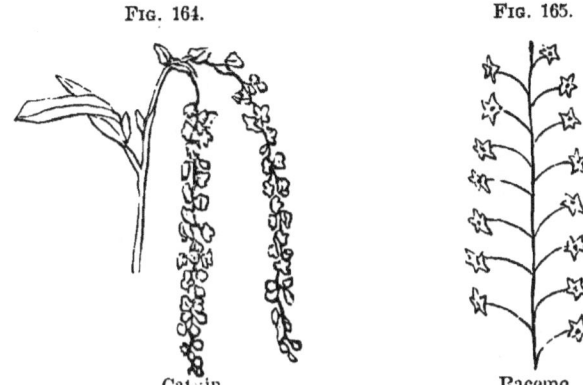

Fig. 164. Catkin. Fig. 165. Raceme.

AMENT, OR CATKIN.—A *spike*, with sessile bracts among its flowers. It grows on trees and shrubs, and falls off after a while.

The RACEME is a flower cluster, where the flowers grow upon pedicels of about equal length along the rachis.

Fig. 166. Glomerule.

A GLOMERULE is formed by nearly sessile clusters of flowers in the axils of opposite leaves.

---

ence of age; but, when this is the case, you must leave the question undecided till you have discovered some other mode of solving it.

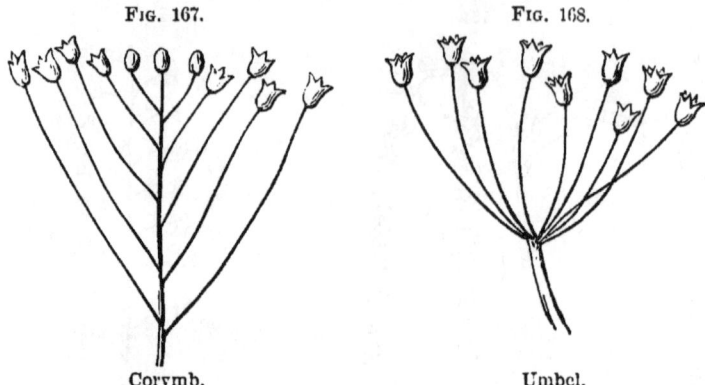

Corymb.   Umbel.

The Corymb is a flower cluster, with a short rachis, the lower pedicels of which are lengthened, so that the cluster is flat at top.

An Umbel has no rachis, and the pedicels are of nearly equal length.

COMPOUND.

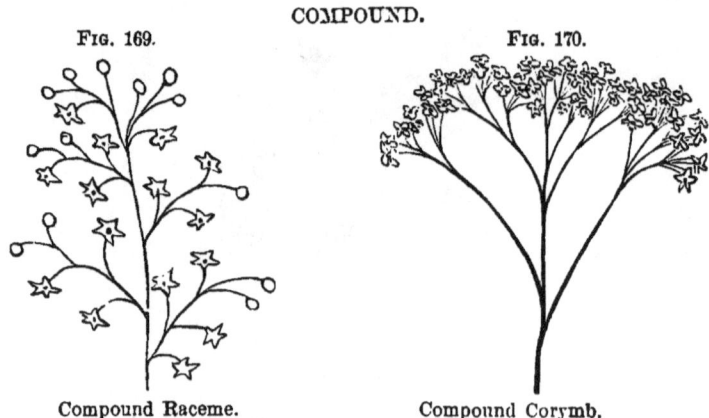

Compound Raceme.   Compound Corymb.

A Compound Raceme, or Panicle, has a long rachis, and the flowers grow upon branches of the pedicels. When such a cluster is thick and cone-shaped, it is called a *Thryse*.

THE INFLORESCENCE.   93

A COMPOUND CORYMB is a corymb with the flowers growing upon branches of the pedicels. Fig. 170.

Fig. 171.

Compound Umbel.

A COMPOUND UMBEL has a second umbel, or umbellet, upon each pedicel.

NOTE.—Most of the clusters pictured in this exercise are represented as without bracts, that differences in their modes of branching may be more easily compared. The pictures represent certain styles of flowering, and each of these styles varies very much in nature. You will find umbels very unlike each other, and very unlike Fig. 170, but still more nearly like that figure than any of the others. And so of panicles, corymbs, &c. Great differences among the clusters of one variety may be occasioned by the presence or absence of bracts, by their groupings, forms, and colors, by the length, stiffness, and even varying positions of peduncles and pedicels, as well as by differences in the form of receptacles. And besides, the various sorts run together in many different ways. You will sometimes find a flower-cluster resembling two different varieties so much that you will have to combine the names of the two in order to characterize it properly; as, for instance, a corymbose panicle, a panicle of heads, or a spicose umbel. When you cannot name the variety, say so, and keep the instance in mind until it becomes clear to you.

94   THE FIRST BOOK OF BOTANY.

FIG. 172.   FIG. 173.   FIG. 174.

Note.—The difference between definite and indefinite flower-clusters is shown above. Fig. 172 represents an indefinite raceme, the growing end of which is surrounded by unopened flowers. In Fig. 174 the reverse is the case; the rachis *ends with a flower*—the oldest flower of the cluster, while at the other end, near the peduncle, the buds have scarcely begun to unfold. This, therefore, is clearly a definite raceme. In Fig. 173 the oldest flowers of the cluster are near the peduncle, the growing end is surrounded by undeveloped buds, and its kind is easily determined.

THE INFLORESCENCE. 95

SCHEDULE TWENTY-ONE, DESCRIBING FIG. 173.

Parts?	Peduncle, Bracts, Rachis, Pedicels, Flowers.
Attitude?	Erect.
Position?	Terminal.
Kind?	Indefinite.
Variety?	Raceme.

LEAF.—Cauline, simple, sessile, exstipulate, feather-veined, serrate, oval-acute.

STEM.—Erect, round, herbaceous.

This is the last inflorescence schedule, and future descriptions of this part of plants will be made without the help of questions. There are some obvious characters of the inflorescence, easily understood and described, that have not been named in the schedule, and, that they may be noted in future descriptions, we call attention to them here.

When many flowers are crowded upon a rachis, or receptacle, the cluster is said to be *dense;* but when they are few and scattering, it is said to be *loose.*

The bracts of a cluster may be very numerous, or they may present peculiarities that a child can easily describe, such, for instance, as relate to shape or color, or they may form an involucre at the base of the cluster, and these points might well be included in a description.

CHAPTER IV.

THE FLOWER.

EXERCISE XXXV.

Parts of the Flower.

Fig. 175 represents one flower—the parts, though separated, stand in their natural relation to each other.

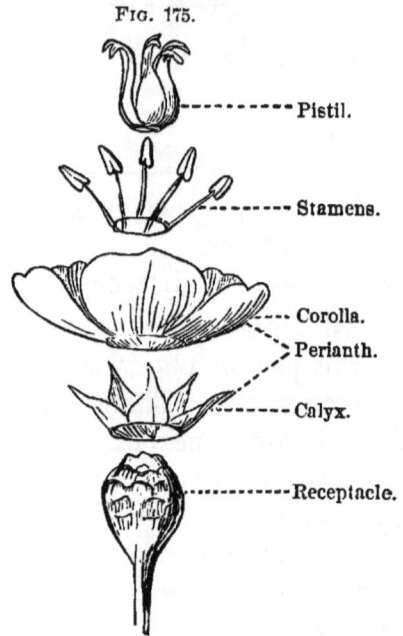

Fig. 175.

Recep'tacle.—The top of the peduncle, more or less swollen, from which the flower grows.

Ca'lyx.—The outer circle of green flower-leaves.

Corol'la.—The inner circle of delicately-colored flower-leaves.

THE FLOWER.

Per'ianth.—A name given to both circles of flower-leaves when they are so nearly alike as not to be separable into calyx and corolla.

Sta'mens.—Slender, thread-like parts next inside the corolla.

Pis'til.—The central part of the flower inside the stamens.

When there is but one whorl of flower-leaves, whatever its color, it is called a calyx.

EXERCISE XXXVI.

Parts of the Calyx.

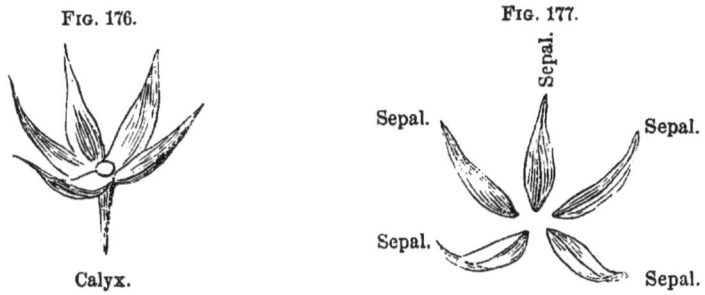

Fig. 176.   Fig. 177.

Calyx.   Sepal. Sepal. Sepal. Sepal.

Se'pal.—One of the leaves of the calyx.

Note.—The first thing in studying the flower is to become acquainted with its leading parts and their names. This is done by comparing numerous specimens with Fig. 175. The pupil is then ready to begin work with the flower schedule. Figs. 177 and 179 are given to assist the pupil in answering the first questions upon it. Write under the question, calyx? the names of the parts that compose the calyx, and under the question, corolla? the names of the parts that compose the corolla. Then count the sepals in your flower, and write their number after the word sepals, in the next column; count also the petals in the corolla, and write their number after the word petals.

EXERCISE XXXVII.

Parts of the Corolla.

FIG. 178. FIG. 179.

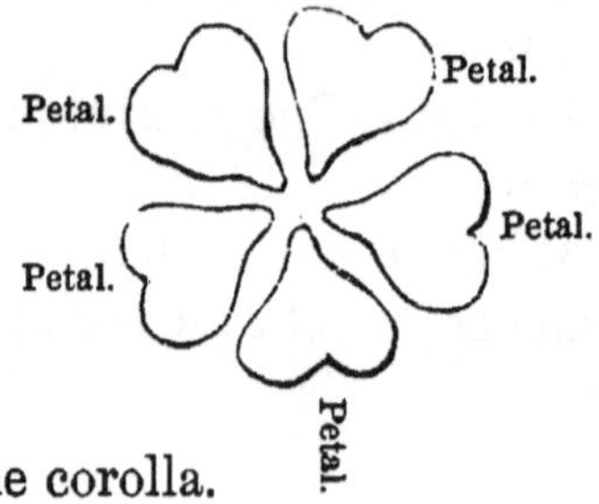

Corolla. Petal. Petal. Petal. Petal. Petal.

PET′AL.—A leaf of the corolla.

FIG. 180.

FIG. 181.

SCHEDULE TWENTY-TWO, DESCRIBING FIG. 178.

Names of Parts.	No.
Calyx ?	
Sepals.	5.
Corolla ?	
Petals.	5.

SCHEDULE TWENTY-THREE, DESCRIBING FIG. 179.

Names of Parts.	No.
Perianth ?	
Leaves.	6.

EXERCISE XXXVIII.

Kinds of Calyx.

Fig. 182.

Polysepalous Calyx.

Fig. 183.

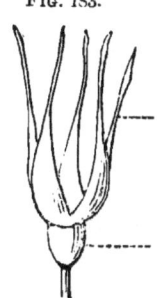

Gamosepalous Calyx.

A Polysep′alous Calyx has its sepals free from each other, so that each one can be pulled off separately.

A Gamosep′alous Calyx has its sepals more or less grown together by their edges, so that, if you pull one, the whole calyx comes off.

Having used schedules twenty-two and twenty-three till the names of the parts that compose the calyx, corolla, and perianth, are firmly associated with the parts themselves, we are now ready to begin their description. Schedule twenty-four shows you where to write what you have to say about them. Observe first whether the sepals of a calyx, the petals of a corolla, or the leaves of a perianth, are grown together or not. Sometimes they cohere so slightly, that close observation is necessary to ascertain it. Be cautious about pronouncing a corolla polypetalous until you have made many observations upon different specimens of it. Do not guess.

You can count the petals of gamopetalous corollas by their marks of cohesion.

EXERCISE XXXIX.

Kinds of Corolla and Perianth.

Fig. 184.

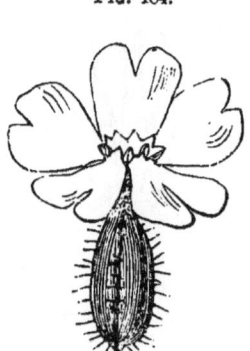

Polypetalous Corolla.

Fig. 185.

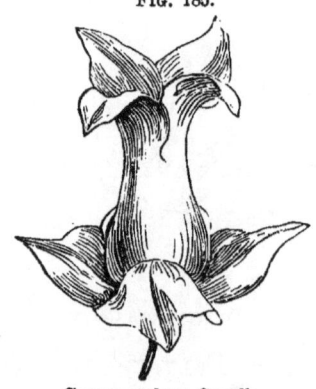

Gamopetalous Corolla.

A Polypet′alous Corolla has its petals free and separate from each other, so that each one can be pulled off without disturbing the others.

A Gamopet′alous Corolla has its petals more or less grown together by their edges, so that if you pull one the whole corolla comes off.

Schedule Twenty-four, describing Fig. 185.

Names of Parts.	No.	Description.
Calyx ?		Gamosepalous.
Sepals.	4.	
Corolla ?		Gamopetalous.
Petals.	4.	

THE FLOWER. 101

A Polyphyl'lous Perianth has its leaves entirely free and separate from each other.

A Gamophyl'lous Perianth has its leaves more or less coherent by their edges.

In the schedule will be seen a space where the forms of sepals and petals should be recorded in the same terms used to describe leaves.

EXERCISE XL.

Regular and Irregular Corollas and Perianths.

Fig. 186.

Fig. 187.

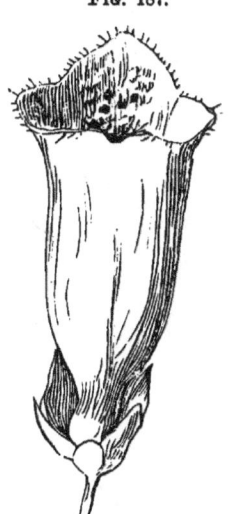

Regular Gamopetalous Corolla.   Irregular Gamopetalous Corolla.

A Regular Calyx, Corolla, or Perianth, has its parts of the same size and shape.

An Irregular Calyx, Corolla, or Perianth, has its parts unlike in size or form.

Fig. 188.

Fig. 189.

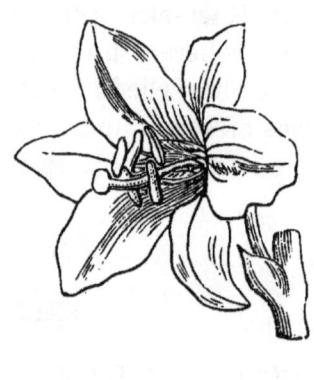

Schedule Twenty-five, describing Fig. 188.

Names of Parts.	No.	Description.
Calyx ? Sepals.	6.	Gamosepalous, irregular.
Corolla ? Petals.	3.	Gamopetalous, irregular.

Schedule Twenty-six, describing Fig. 189.

Names of Parts.	No.	Description.
Perianth ? Leaves.	6.	Polyphyllous, regular.

THE FLOWER. 103

EXERCISE XLI.

Parts of Stamens.

Fig. 190.

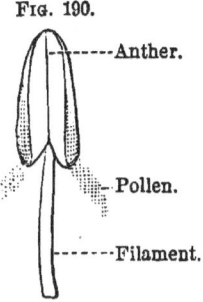

Fil'ament.—The stem-like part of a stamen.
An'ther.—The thickened oblong head of a filament.
Pol'len.—The dust, or powder, seen upon the anther.

Schedule twenty-seven has added to it the new question, Stamens? Write underneath it the name of the parts that compose a stamen of your flower. Count the number of stamens, and write it down, unless they are too numerous, when you will use the character ∞, signifying many. Write free, when they are so; and coherent, when they are grown together.

When the filament is absent, write sessile after anther. To describe the filaments, observe whether they are long or short, slender or thick, flat or round, free or grown together.

Observe whether the anthers are one-lobed or two-lobed, that is, whether they are in two parts or pieces; and note also whether they are oblong, round, curved, straight, large or small, longer or shorter than the filaments, free or grown together.

Fig. 191.

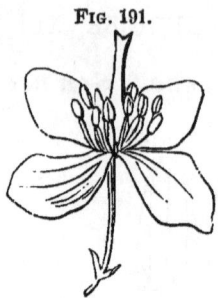

Schedule Twenty-seven, describing Fig. 191.

Names of Parts.	No.	Description.
Calyx?		Gamosepalous, regular.
Sepals.	2.	
Corolla?		Polypetalous, regular.
Petals.	4.	Obovate, Spreading.
Stamens?	∞	
Filament.		Slender.
Anther.		Two-celled, Oblong.

Note.—Our descriptions of pictured flowers are necessarily imperfect, because the pictures are themselves imperfect. As the pollen is not represented in Fig. 191, it is, of course, omitted from the schedule. We can say nothing, in a book, of the color or size of specimens; yet the plan of working is clearly illustrated, and pupils will not find it difficult, at this stage, to add such points without the guidance of a pattern schedule.

THE FLOWER.

EXERCISE XLII.

Parts of the Pistil.

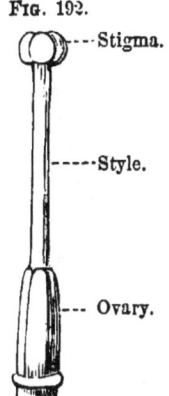

Fig. 192.
Stigma.
Style.
Ovary.

O'vary.—The lowest part of the pistil, containing the seeds.

Style. — The slender stem-like part of the pistil next above the ovary.

Stig'ma. — The top of the pistil.

EXERCISE XLIII.

Parts of the Ovary.

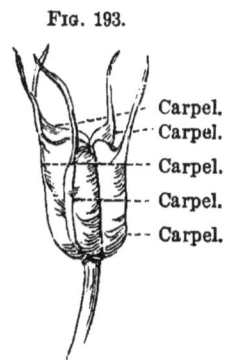

Fig. 193.
Carpel.
Carpel.
Carpel.
Carpel.
Carpel.

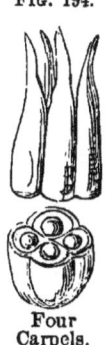

Fig. 194.
Four Carpels.

Car'pel.—One of the divisions, or cells, of the ovary.

106 THE FIRST BOOK OF BOTANY.

Fig. 195. Fig. 196. Fig. 197.

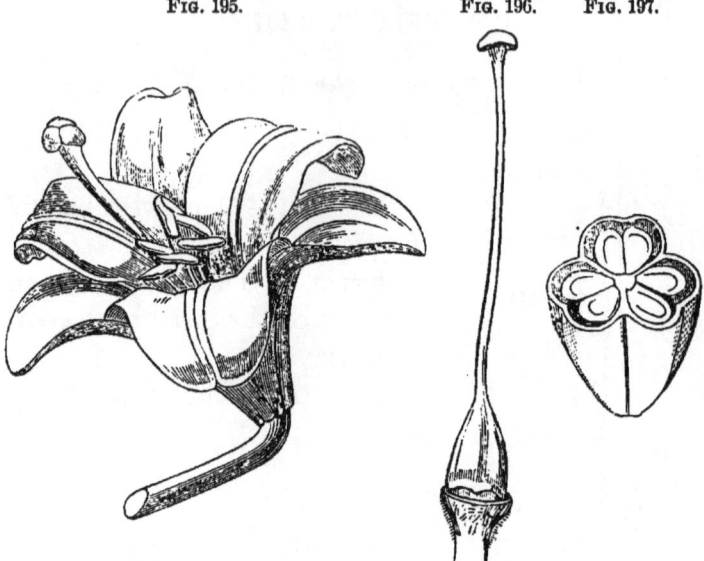

Schedule Twenty-eight, describing Fig. 195.

Names of Parts.	No.	Description.
Perianth ? *Leaves.*	6.	*Polyphyllous, regular.*
Stamens ? *Filament.* *Anther.*	6.	*Free.* *Slender.* *Oblong.*
Pistil ? *Carpels.* *Style.* *Stigma.*	3.	*A single column.* *Three-lobed.*

THE FLOWER.

The question Pistil? is now added to the schedule, and is to be answered in the same way as the questions Perianth? and Stamens? First write the name of its parts underneath, and then find out, if you can, the number of carpels that compose the ovary. It is sometimes quite difficult to do this, but it is well always to make the effort. When the carpels cannot be distinguished, you determine their number by counting the styles, and, if these are grown smoothly together, then count the lobes of the stigma. It is very seldom that this part of the pistil is so coherent that the lines of union are invisible. You can often, in this way, find out the number of carpels in a pistil, when every other means fails. In describing the various forms of style no new terms are needed.

EXERCISE XLIV.

Parts of the Petals.

Fig. 198.

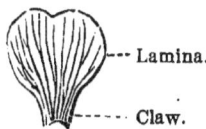

--- Lamina.
----- Claw.

Fig. 199.

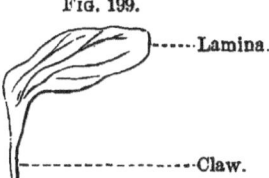

------ Lamina.
---------------- Claw.

Lam′ina.—The upper, and usually the broadest and thinnest, part of a petal.

Claw.—The lower part of a petal, which attaches it to the receptacle.

EXERCISE XLV.

Kinds of Regular Polypetalous Corollas.

Fig. 200.

Cruciferous.

Fig. 201.

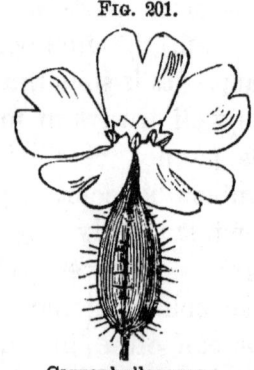

Caryophyllaceous.

A CRUCIF'EROUS COROLLA has four petals growing in the shape of a cross.

A CARYOPHYLLA'CEOUS COROLLA has five petals, having each a long, slender claw, and a spreading blade.

Fig. 202.

Rosaceous.

Fig. 203.

Liliaceous.

A ROSA'CEOUS COROLLA has five petals, with spreading lamina and short claw.

A LILIA'CEOUS PERIANTH has six leaves, bending away something like a bell.

EXERCISE XLVI.

Kinds of Irregular Polypetalous Corolla.

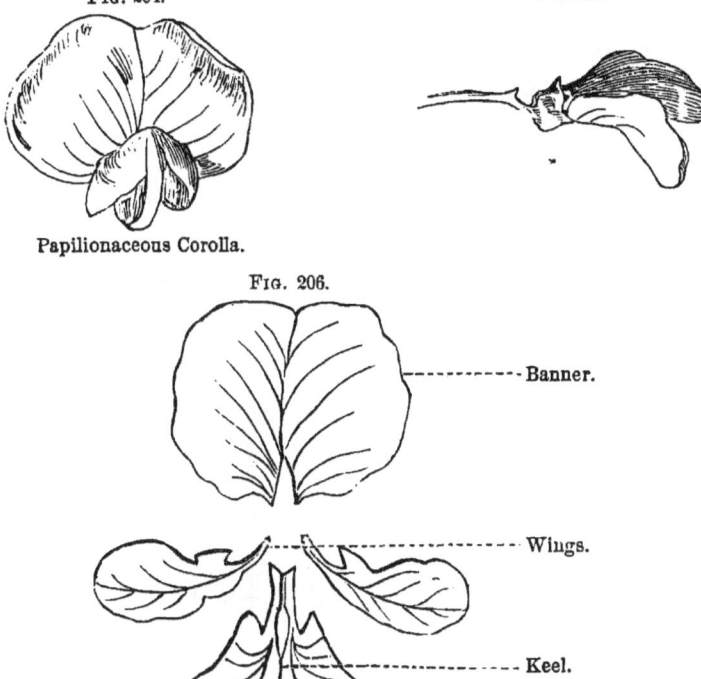

Fig. 204. Papilionaceous Corolla.

Fig. 205.

Fig. 206.—Banner. Wings. Keel.

The PAPILIONA'CEOUS COROLLA has five dissimilar petals, arranged like Fig. 204. The one nearest the stem (the upper, Fig. 206) is called the *banner;* the two side ones are called *wings*, and the lower one the *keel*.

NOTE.—Learn to distinguish the banner, wings, and keel of papilionaceous corollas, and note the differences of their forms in different kinds of flowers. You can write such observations upon the back of the schedule.

FIG. 207.   FIG. 208.

FIG. 209.

There are many other varieties of polypetalous irregular corollas which are described generally as *anomalous*. Fig. 207 is a common form of anomalous corolla. There is an interesting tribe of plants known as orchids, which present many anomalous forms of corolla; Fig. 208 is an example, Fig. 209 being a separate flower from the same plant. Anomalous flowers should be further described as polypetalous or gamopetalous, for they occur among both these forms.

THE FLOWER.

FIG. 210.   FIG. 211.   FIG. 212.

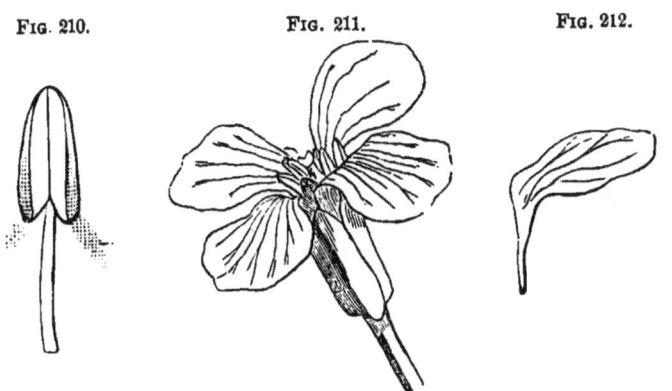

SCHEDULE TWENTY-NINE, DESCRIBING FIG. 211.

Names of Parts.	No.	Description.
Calyx ? Sepals.	 4.	Polysepalous, regular. Oval.
Corolla ? Petals.	 4.	Cruciform. Claw, long, Limb, spreading.
Stamens ? Anther. Filament. Pollen.	6.	 Oblong. Short as anther.
Pistil ?		

EXERCISE XLVII.

Parts of a Gamopetalous Corolla.

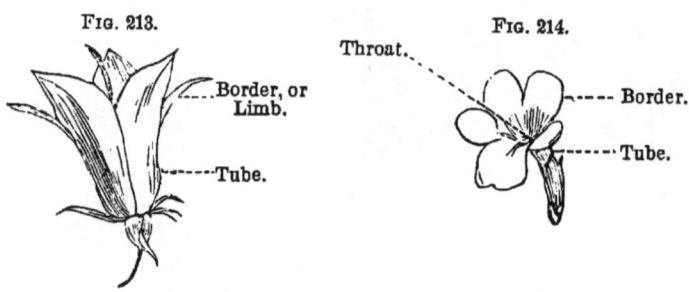

Tube.—That part of the corolla, whether long or short, in which the petals are united together.

Limb, or Border.—The upper part of the corolla, where the petals are not united.

Throat.—The opening of the tube.

Corolla Tubes may be long or short, slender or swollen, tapering or cylindrical, or with a pouch, or sack, on one side.

The Limb may be narrow or broad, erect or spreading; and,

The Throat is either open or constricted, hairy or smooth.

Note these features in describing gamopetalous corollas.

Note.—The last exercises of this chapter introduce twenty or thirty new terms, expressive of as many different ideas of form. In learning the precise word for each form, proceed very slowly from exercise to exercise, searching constantly for illustrative specimens. Learn the names of the parts of a petal and of a gamopetalous corolla. Let time be taken to examine all the flowers that can be found, comparing their corollas with the pictures, fixing, for each flower, upon the picture it most nearly resembles.

THE FLOWER. 113

EXERCISE XLVIII.

Kinds of Regular Gamopetalous Corollas.

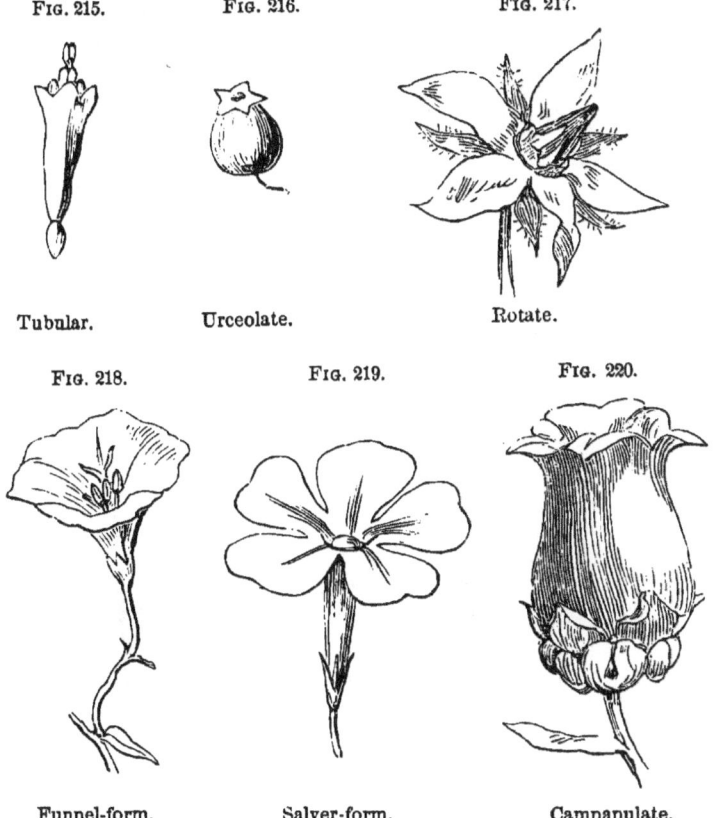

Fig. 215. Tubular. Fig. 216. Urceolate. Fig. 217. Rotate.
Fig. 218. Funnel-form. Fig. 219. Salver-form. Fig. 220. Campanulate.

Tu′bular.—A tubular corolla is one in which the tube spreads little or none at the border. Fig. 215.

Ur′ceolate.— A corolla is urceolate when the tube is swollen in the middle, with a narrow opening like an urn, as in Fig. 216.

Ro′tate, or Wheel-shaped Corollas have a short tube and flat, spreading border. Fig. 217.

Fun'nel-form.—When the corolla-tube is small below, and enlarges gradually to the border, as in Fig. 218.

Sal'ver-form.—When the long, slender tube of a corolla ends abruptly in a flat spreading border, as seen in Fig. 219.

Campan'ulate.—Bell-shaped corollas are said to be campanulate. Fig. 220.

EXERCISE XLIX.

Irregular Gamopetalous Corollas.

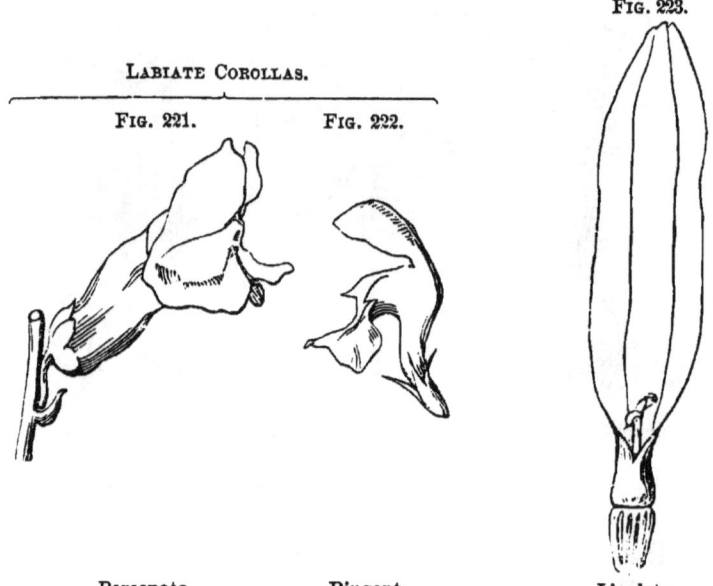

Labiate Corollas.

Fig. 221.   Fig. 222.   Fig. 223.

Personate.   Ringent.   Ligulate.

La'biate.—In labiate corollas the limb has the appearance of lips ; Figs. 221 and 222. Labiate corollas are of two kinds, personate and ringent.

THE FLOWER. 115

Per'sonate.—With the throat open. Fig. 221.
Rin'gent.—With the throat closed. Fig. 222.

A Lig'ulate, or strap-shaped, corolla, is one which appears as if it were formed by the splitting of the tube on one side. Fig. 223.

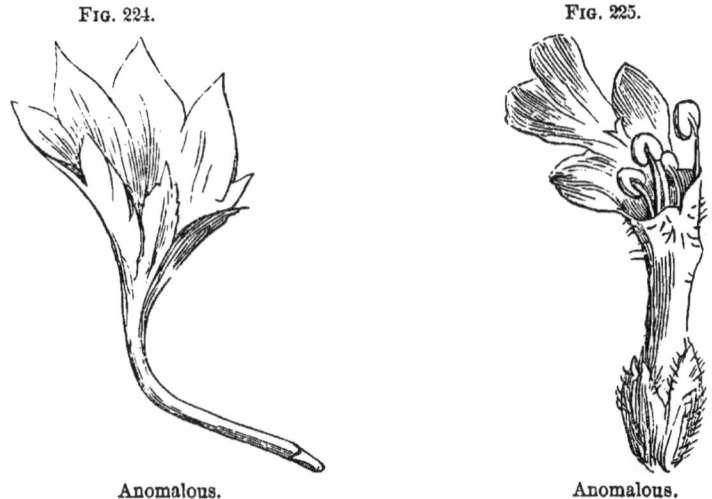

Fig. 224.   Fig. 225.

Anomalous.   Anomalous.

Anom'alous.—All other irregular gamopetalous corollas, as Figs. 224 and 225, are called *anomalous*.

In describing corollas, the terms cruciferous, lillaceous, tubular, etc., may now be used in place of polypetalous, gamopetalous, regular and irregular, as the new terms include these characters, along with others, more limited and special. To say, for example, that a corolla is *cruciferous*, is to say that it is polypetalous and regular, and also to state the number and position of its petals. To say that a corolla is *strap-shaped*, is the same as saying that it is gamopetalous and irregular as well as what particular form it has.

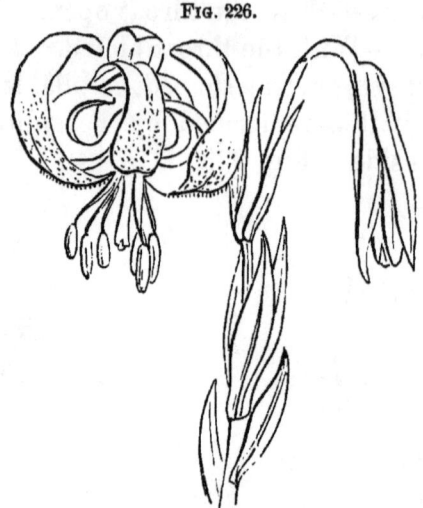

Fig. 226.

Schedule Thirty, describing Fig. 226.

Names of Parts.	No.	Description.
Perianth ?		Liliaceous.
Leaves.	6.	Lanceolate, Recurved.
Stamens ?	6.	
Filament.		Long, Slender.
Anther.		Two-celled, Oblong.
Pistil ?		
Carpels.	3.	
Style.		Single, Smooth.
Stigma.		Three-cleft.

THE FLOWER.   117

FIG. 227.

SCHEDULE THIRTY-ONE, DESCRIBING FIG. 227.

Names of Parts.	No.	Description.
Calyx ? Sepals.	5.	Gamosepalous.
Corolla ? Petals.	5.	Salver-form.
Stamens ? Filament. Anther.	5.	
Pistil ? Carpels. Style. Stigma.		

EXERCISE L.

Crowns, Spurs, and Nectaries.

Fig. 228.

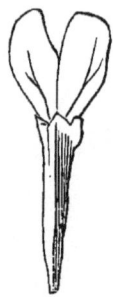

Fig. 229.

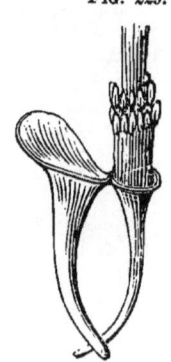

The CORONA, or CROWN, is a scale-like structure (Fig. 228) on the inner surface of corollas, at the summit of the claw, or tube.

A SPUR is a tubular prolongation of a petal or sepal. Fig. 229.

Fig. 230.

Nectary.

Fig. 231.

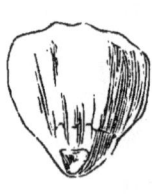

Nectary.

A NECTARY is a little gland on the claw of a petal that secretes a sugary liquid. In Fig. 230 these glands are naked, while in Fig. 231 the little gland is covered by a scale.

THE FLOWER.

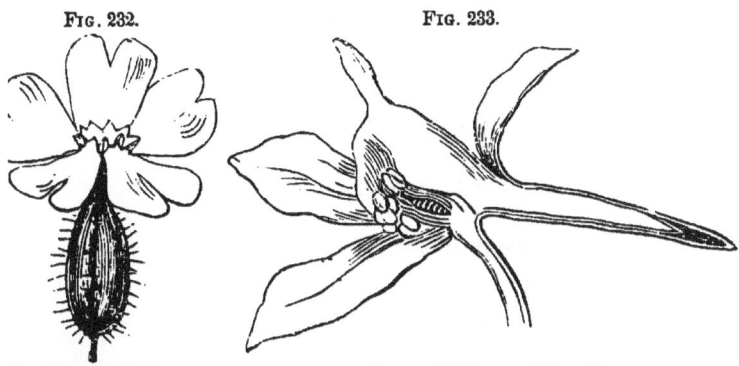

Corolla with Crown.   Spurred Calyx and Corolla.

The statement proper in describing the corolla (Fig. 232) is as follows:

| Corolla? | | *Caryophyllaceous.* |
| Sepals. | 5. | *Limb, obcordate; crown at base.* |

We describe a calyx and corolla like that shown (Fig. 233) as follows:

Calyx?		
Sepals.	5.	*Upper one spurred.*
Corolla?		*Polypetalous, anomalous.*
Petals.	4.	*Upper one with spur, prolonged into calyx spur.*

CHAPTER V.

THE ROOT.

EXERCISE LI.

Tap-Roots and Fibrous Roots.

There are two classes of roots, called tap-roots and fibrous roots, which differ from each other in the way shown in Figs. 234 and 235.

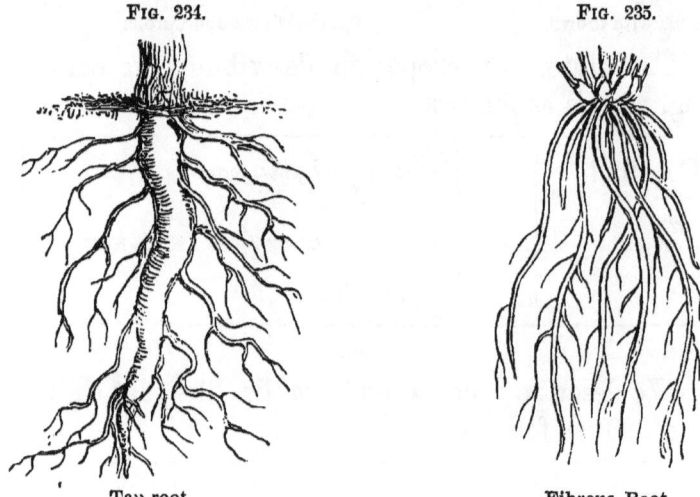

Fig. 234. Tap-root.

Fig. 235. Fibrous Root.

Fig. 234 represents a Tap-root, which is seen to be simply a continuation of the stem downward.

In Fig. 235 the stem is not continued downward as a tap-root, but sends off rootlets or fibres at the outset. It is hence called a Fi′brous Root.

EXERCISE LII.

Kinds of Tap-Root.

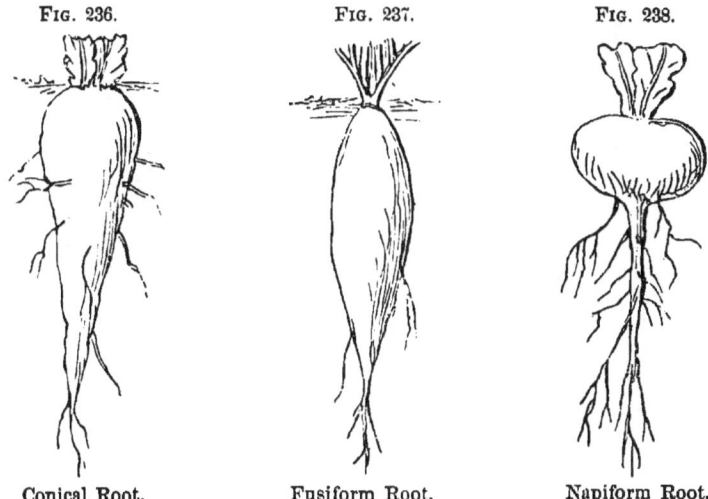

Fig. 236.   Fig. 237.   Fig. 238.
Conical Root.   Fusiform Root.   Napiform Root.

Con'ical Roots are tap-roots, which taper gradually downward, and so are shaped like a cone. Fig. 236.

Fu'siform, or Spindle-shaped Roots, are tap-roots enlarged in the middle of their length, and tapering toward both ends. Fig. 237.

A Nap'iform, or Turnip-shaped Root (Fig. 238), is a tap-root, more or less globular in form.

The kinds of tap-root illustrated in this exercise are equally continuations of the stem, with that shown in Fig. 234. By reference to Fig. 116, it will be seen that these stems are made up of nodes, and are just as really stems as those in which the intervals between the nodes are considerable.

EXERCISE LIII.

Kinds of Fibrous Roots.

Fig. 239.

Fig. 240.

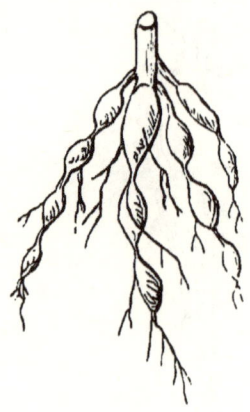

Moniliform Root.

Fig. 241.

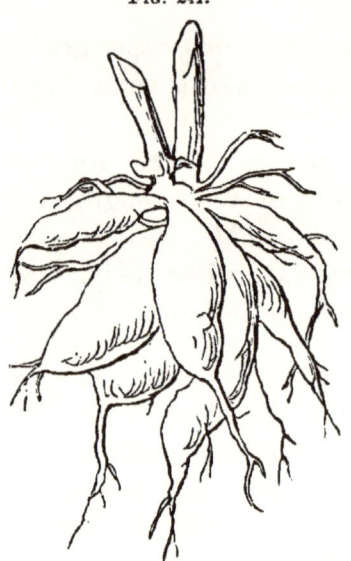

Fig. 242.

Fasciculated Root.

Tubercular Root.

In MONIL'IFORM ROOTS (Fig. 240) some of the fibres have numerous small swellings, that succeed each other so as to look like a string of beads.

In FASCIC'ULATED ROOTS (Fig. 241) the fibres become swollen along their length, and look like a bundle of fusiform roots.

When some of the rootlets of fibrous roots become fleshy and enlarged, taking the form shown in Fig. 242, they are called TUBERCULAR ROOTS.

NOTE.—It is not difficult to see that the moniliform root is only a fibrous root, in which regular portions of the fibres have become swollen. When all these swellings unite in one continuous enlargement, we have a fasciculated root (Fig. 241). When the swellings are shortened and globular (Fig. 242), we name them tubercular roots, but their resemblance to the fibrous root is still apparent.

The questions about roots suggested by this chapter are, first, is the specimen in hand a tap or fibrous root? The answer may not always be easy, but the pupil will exercise his best judgment upon it. If it be fibrous, however, say so; if any modification of fibrous, say which, and similarly if the kind be a tap-root. For aid in describing roots, we must refer pupils to the exercises in plant description, which follow.

There is usually a certain balance between the size of the root and stem of a plant; but sometimes the root is very small compared with the stem and branches, and sometimes it is large. Roots may also be loosely attached to the soil or firmly planted therein; they may be spreading near the surface, or may grow directly downward, and such facts are worthy of note in root descriptions.

EXAMPLES IN PLANT DESCRIPTION,

ILLUSTRATIVE OF THE FOREGOING EXERCISES.

Fig. 242.

Description of Fig. 242.

ROOTS fibrous.

LEAVES radical, petiolate, exstipulate, palmate-veined, entire, acutely three-lobed; base cordate, surface hairy. *Bracts* hairy, in a whorl of three near the flower.

INFLORESCENCE solitary, on a slender hairy scape.*

FLOWER. CALYX; *sepals* 8–12, oblong spreading: COROLLA none: STAMENS many; *filaments* threadlike; *anthers* oval, two-celled: PISTIL; *carpels* many; *style* very short; *stigma* continued down the inner face of the style.

---

* Scape, a peduncle which arises from an underground stem.

Fig. 243.

Description of Fig. 243.

Roots fibrous.

STEM a scaly bulb.

LEAVES radical, petiolate, exstipulate, digitately three-fingered; *leaflets* sessile, feather-veined, entire, obcordate; *petiole* long, slender.

INFLORESCENCE a loose terminal umbel.

FLOWER. CALYX; *sepals* 5, polysepalous: COROLLA; *petals* 5, regular, polypetalous, obovate, much larger than the sepals: STAMENS 10, of unequal length, hairy; *filaments* awl-shaped, flattened below, grown together; *anthers* short, oval, two-celled: PISTIL: *ovary* ovoid, of 5 united *carpels; styles* free, hairy; *stigmas* enlarged, rounded.

FIG. 244.

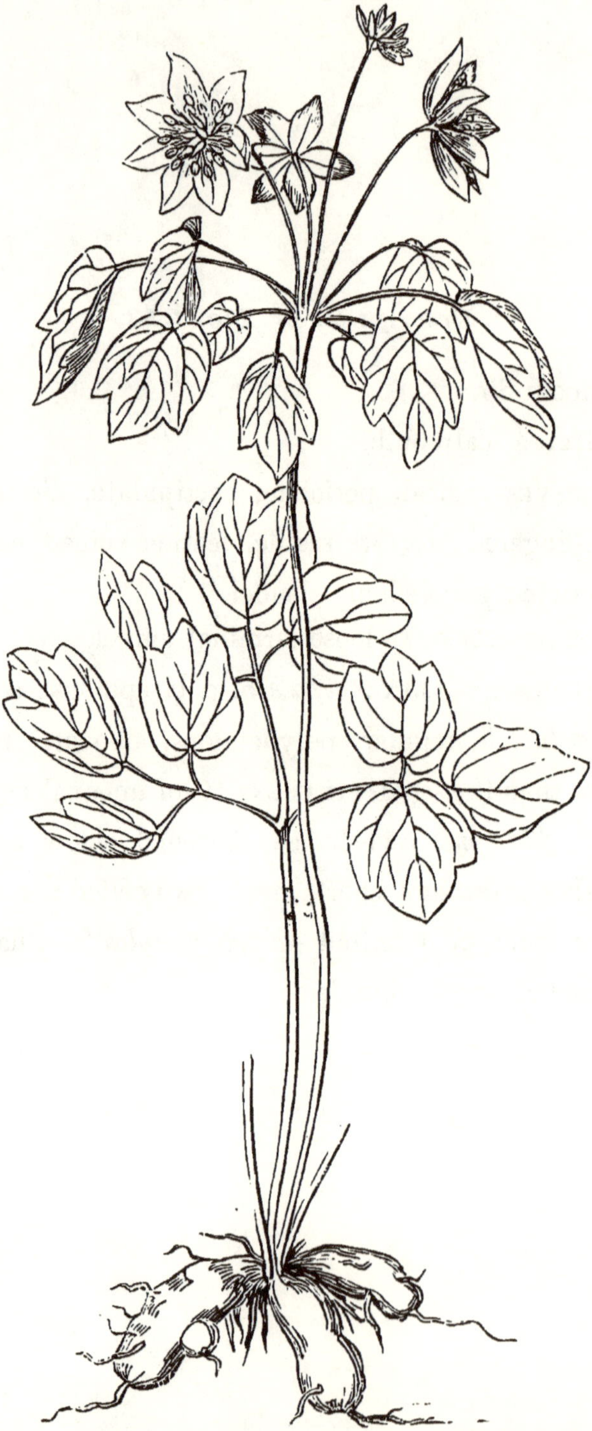

Description of Fig. 244.

ROOTS fasciculated.

STEM erect, round, slender, herbaceous.

LEAVES radical and cauline. Radical leaves twice ternately three-fingered ; *leaflets* petiolulate, palmate-veined, entire, three-lobed at the end, subcordate ; *petiole* long and slender. Cauline leaves numerous, simple, petiolate, exstipulate, formed like the leaflets of the radical leaves, placed in a whorl at the base of the inflorescence.

INFLORESCENCE a loose terminal umbel.

FLOWER. CALYX ; *sepals* 6–8, spreading, polysepalous, regular : COROLLA ; *petals* none : STAMENS many ; *filaments* thread-like ; *anthers* two-celled : PISTIL ; *carpels* many.

FIG. 245.

Description of Fig. 245.

Roots branching tap.

STEM erect, slender, herbaceous, round, hairy.

LEAVES cauline, opposite, simple, sessile, exstipulate, entire, ovate-acute.

INFLORESCENCE clustered, terminal, umbellate.

FLOWER. CALYX; *sepals* 5 : COROLLA; *petals* 5, obcordate, spreading : STAMENS 10 ; *filaments* thread-like ; *anthers* oval, two-celled : PISTIL ; *ovary* ovoid, consisting of five united carpels; *styles* short, free; *stigma* along the inner face of the style.

Fig 246.

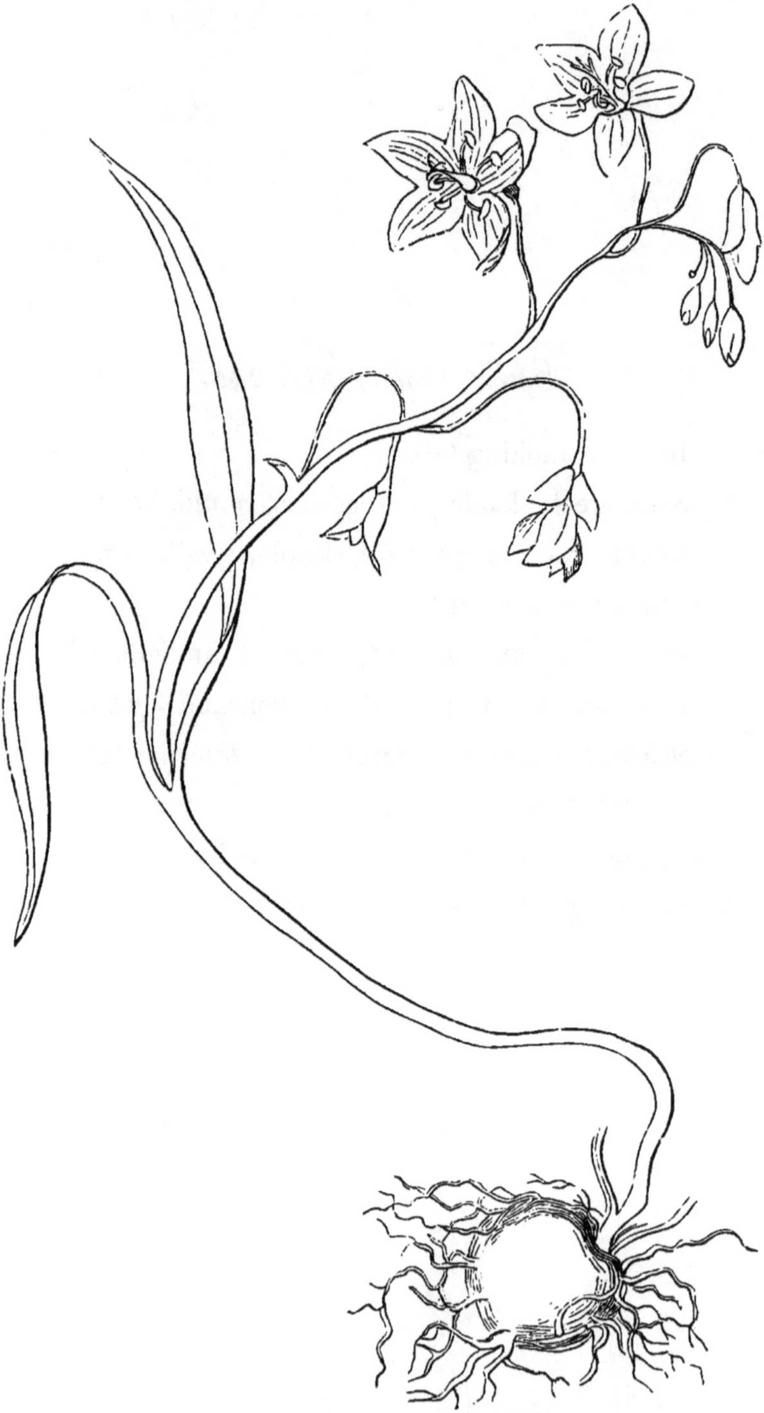

Description of Fig. 246.

ROOTS tuberous.

STEM smooth, low, weak, slender, herbaceous, round.

LEAVES cauline, opposite, a single pair, sessile, exstipulate, feather-veined, entire, lanceolate.

INFLORESCENCE a loose definite raceme.

FLOWER. CALYX ; *sepals* 2, polysepalous, regular : COROLLA ; *petals* 5, polypetalous (or slightly coherent at the short claws), spreading ; *stamens* 5 ; *filaments* threadlike ; *anthers* oval : PISTIL ; carpels 3 ; *style* slender, three-cleft ; *stigma* along the inner side of the three-cleft style.

Fig. 247.

Description of Fig. 247.

ROOTS fibrous, matted, somewhat spreading.

STEM of scaly nodes, internodes none.

LEAVES radical, simple, exstipulate, peltately palmate-veined, wavy, deeply two-lobed, shut sinus at base; *petiole* long, round, rather erect.

INFLORESCENCE solitary, on a smooth, naked scape.

FLOWER. CALYX; *sepals* 4, polysepalous, oblong: COROLLA; *petals* 8, polypetalous, regular, oblong, spreading: STAMENS 8; *filaments* threadlike, shorter than anther; *anthers* two-celled, oblong: PISTIL; carpels 2; *style* short; *stigma* spreading, two-lobed.

Fig. 248.

Description of Fig. 248.

Roots fibrous, growing from the entire under-side of the stem.

Stem creeping below the ground.

Leaf radical, petiolate, exstipulate, wavy-dentate, palmate-veined, slightly reniform, obtusely seven-lobed, sinuses rounded, nearly closed; *petiole* half-round, channelled.

Inflorescence solitary, on a smooth, slender scape.

Flower. Calyx; *sepals* 2, ovate, regular: corolla; *petals* 8, polypetalous, regular, obovate-oblong, spreading: stamens many, shorter than the petals; *filaments* short, threadlike; *anthers* oblong, two-celled: pistil; ovary oblong, of two carpels; *styles* united in a column; *stigma* two-lobed.

Fig. 249.

Description of Fig. 249.

ROOTS fasciculated.

STEM slender, weak, round, herbaceous, hairy.

LEAVES radical and cauline. Radical leaves, petiolate, exstipulate, entire, deeply twice ternately lobed; *petioles* long, hairy. Cauline leaves sessile alternate, shaped like the radical leaves, but much smaller.

INFLORESCENCE solitary, terminal.

FLOWER. CALYX ; *sepals* 5, polysepalous, regular, spreading : COROLLA ; *petals* 5, polypetalous, regular, oval, spreading : STAMENS many ; *filaments* threadlike ; *anthers* short, two-celled : PISTIL ; *carpels* many ; *styles* very short or absent ; *stigma* inner and upper part of carpel or style.

Fig. 250.

Description of Fig. 250.

ROOTS moniliform.

STEM erect, slender, herbaceous, round.

LEAVES radical and cauline, ternately compound. Cauline leaves alternate ; *leaflets* lobed ; *petioles* spreading at base.

INFLORESCENCE solitary, terminal.

FLOWER. CALYX ; *sepals* 5, polysepalous, regular, spreading, ovate : COROLLA ; *petals*, none : STAMENS numerous ; *filaments* threadlike ; *anthers* oblong : PISTIL ; *carpels* many ; *stigma* sessile on the upper, inner face of carpel.

LEAF SCHEDULES.

SCHEDULE ONE.
See Page 19, *Exercises I., II., III., IV., and V.*

Parts?	
Venation?	

SCHEDULE TWO.
See Page 27, *Exercise VI.*

Parts?	
Venation?	
Margin?	

SCHEDULE THREE.
See Page 30, *Exercise VII.*

Parts?	
Venation?	
Margin?	
Base?	

Schedule Four.

See Page 32, *Exercise VIII.*

Parts ?	
Venation ?	
Margin ?	
Base ?	
Apex ?	

Schedule Five.

See Page 34, *Exercise IX.*

Parts ?	
Venation ?	
Margin ?	
Base ?	
Apex ?	
Lobes ?	

SCHEDULE SIX.

See Page 36, Exercise X.

Parts ?	
Venation ?	
Margin ?	
Base ?	
Apex ?	
Lobes ?	
Sinuses ?	

SCHEDULE SEVEN.

See Page 38, Exercises XI. and XII.

Kind ?	
Venation ?	
Margin ?	
Base ?	

Schedule Seven.—(*Continued.*)

Apex ?	
Lobes ?	
Sinuses ?	
Shape ?	

Schedule Eight.

See Page 44, *Exercises XIII. and XIV.*

Kind ?	
Venation ?	
Margin ?	
Base ?	
Apex ?	
Lobes ?	
Sinuses ?	
Shape ?	

LEAF SCHEDULES. 147

SCHEDULE EIGHT.—(*Continued.*)

Petiole ?	
Color ?	
Surface ?	

SCHEDULE NINE.

See Page 50, *Exercise XV.*

Parts ?	
No. of Leaflets ?	

SCHEDULE TEN.

See Page 51, *Exercise XVI.*

Parts ?	
No. of Leaflets ?	
Kind ?	

Schedule Eleven.

See Page 56, *Exercises XVII. and XVIII.*

Parts?	
No. Leaflets?	
Kind?	
Variety?	

STEM SCHEDULES.

Schedule Twelve.

See Page 63, *Exercises XXI. and XXII.*

Parts?	
Appendages?	

Leaf.—

STEM SCHEDULES.

Schedule Thirteen.
See Page 65, Exercise XXIII.

Appendages?	
Leaf-position?	

LEAF.—

Schedule Fifteen.
See Page 68, Exercise XXIV.

Appendages?	
Leaf-position?	
Leaf-arrangement?	

LEAF.—

Schedule Sixteen.
See Page 70, Exercise XXV.

Appendages?	
Leaf-position?	
Leaf-arrangement?	
Shape?	

LEAF.—

Schedule Seventeen.
See Page 74, Exercise XXVI.

Appendages?	
Leaf-position?	
Leaf-arrangement?	
Shape?	
Attitude?	

Leaf.—

Schedule Eighteen.
See Page 75, Exercise XXVII.

Appendages?	
Leaf-position?	
Leaf-arrangement?	
Shape?	
Attitude?	
Color?	

SCHEDULE EIGHTEEN.—(*Continued.*)

Surface?	
Size?	
Structure?	

LEAF.—

INFLORESCENCE SCHEDULES.

SCHEDULE EIGHTEEN.
See Page 81, *Exercises* **XXVIII.**, **XXIX.**, *and* **XXX.**

Parts?	
Attitude?	

LEAF.—

STEM.—

INFLORESCENCE SCHEDULES.

Schedule Nineteen.
See Page 86, *Exercises **XXXI**. and **XXXII**.*

Parts ?	
Attitude ?	
Position ?	

Leaf.—

Stem.—

Schedule Twenty.
See Page 89, *Exercise **XXXIII**.*

Parts ?	
Attitude ?	
Position ?	
Kind ?	

Leaf.—

Stem.—

Schedule Twenty-one.

See Page 95, *Exercise XXXIV.*

Parts?	
Attitude?	
Position?	
Kind?	
Variety?	

Leaf.—

Stem.—

FLOWER SCHEDULES.

Schedule Twenty-two.

See Page 98, Exercises XXXV., XXXVI., and XXXVII.

Names of Parts.	No.	
Calyx ?		
Corolla ?		

Leaf.—

Stem.—

Inflorescence.—

Schedule Twenty-three.

See Page 98, Exercises XXXV., XXXVI., and XXXVII.

Names of Parts.	No.	
Perianth ?		

Leaf.—

Stem.—

Inflorescence.—

FLOWER SCHEDULES. 155

Schedule Twenty-four.
See Page 100, *Exercises* **XXXVIII.** *and* **XXXIX.**

Names of Parts.	No.	Description.
Calyx ?		
Corolla ?		

Leaf.—

Stem.—

Inflorescence.—

Schedule Twenty-five.
See Page 102, *Exercise* **XL.**

Names of Parts.	No.	Description.
Calyx ?		
Corolla ?		

Leaf.—

Stem.—

Inflorescence.—

Schedule Twenty-six.

See Page 102, *Exercise XL.*

Names of Parts.	No.	Description.
Perianth?		

LEAF.—

STEM.—

INFLORESCENCE.—

Schedule Twenty-seven.

See Page 104, *Exercise XLI.*

Names of Parts.	No.	Description.
Calyx?		
Corolla?		
Stamens?		

LEAF.—

STEM.—

INFLORESCENCE.—

Schedule Twenty-eight.

See Page 106, *Exercises XLII. and XLIII.*

Names of Parts.	No.	Description.
Calyx ?		
Corolla ?		
Stamens ?		
Pistil ?		

Leaf.—

Stem.—

Inflorescence.—

Note.—This is the last form of schedule in the book. As the pupil passes on from exercise to exercise, he will be enabled to add one feature after another to his descriptions; but the mode of inserting these new points will not make any change in the form of the schedule.

THE EDUCATIONAL CLAIMS OF BOTANY.

It has been stated in the preface that the present work is the outgrowth of a desire to gain certain advantages in general mental culture, which can be only obtained by making Nature a more direct and prominent object of study in primary education than is now done. I have thought it desirable to present the reasons which have led to its preparation more fully than would be suitable in an introduction, and therefore place them at the close of the work.

The subject of mind has various aspects; that in which the teacher is chiefly concerned is its aspect of *growth*. I propose to consider the subject from this point of view; to state, first, some of the essential conditions of mental unfolding; then to show in what respects the prevailing school culture fails to conform to them; and, lastly, to point out how the subject of Botany, when properly pursued, is eminently suited to develop those forms of mental activity, the neglect of which is now the fundamental deficiency of popular education.

Mind is a manifestation of life; and mental growth is dependent upon bodily growth. In fact, these operations not only proceed together, but are governed by the same laws. As body, however, is something more tangible and definite than mind, and as material changes are more easily apprehended than mental changes, it will be desirable to glance first at what takes place in the growth of the body.

I.—HOW THE BODY GROWS.

All living beings commence as germs. The germ is a little portion of matter that is uniform throughout, and is hence said to be *homogeneous*.*

---

* In the following statement two or three words will occur with which

The beginning of growth is a change in the germ, by which it is separated into unlike parts. One portion becomes different from the rest, or is *differentiated* from it; and then it is itself still further changed or differentiated into more unlike parts. In this way, from the diffused uniform mass, various tissues, structures, and organs gradually arise, which, in the course of growth, constantly become more diverse, complex, and *heterogeneous*. But, accompanying these changes, there is also a tendency to *unity*. It is by the assimilation of like with like that differences arise. Nourishment is drawn in from without, and then each part attracts to itself the particles that are like itself. Bone material is incorporated with bone, and nerve material with nerve; so that each different part arises from the grouping together of similar constituents. This tendency to unity, by which each part is produced, and by which all the parts are wrought together into a mutually dependent whole, is termed *integration;* and the combined operations by which development is carried on constitute what is now known as *Evolution.*

At birth, bodily development has been carried so far that the infant is capable of leading an independent life. Mental growth commences when the little creature begins to be acted upon by *external agencies*. An already-growing mechanism takes on a new kind of action in new circumstances, and body and mind now grow together. The development of mind depends upon certain properties of nervous matter by which it is capable of receiving, retaining, and combining impressions. An organism has been thus prepared, upon which the surrounding universe takes effect, and the growth of mind consists in the development of an internal consciousness in correspondence to the external order of the world.

II.—HOW THE MIND GROWS.

At birth we say the infant *knows* nothing; that is, it recognizes *no thing*. Though the senses produce perfect impres-

---

some readers may be unfamiliar. But more precise thoughts require more precise terms to mark them; and, as these terms are now established, their use here is admissible as well as advantageous.

sions from the first, yet these impressions are not distinguished from each other. This vague, indefinite, homogeneous sensibility or feeling may be called the germ-state of mind. As bodily growth begins in a change of the material germ, so mental growth begins in a change of feeling. This change of feeling is due to a change of external impressions upon the infant organism. Were there no changes of impression upon us, there could never be changes of feeling within us, and *knowing* would be impossible. If, for example, there were never an alteration of temperature, and a resulting change of impressions upon the nerves, we should be forever prevented from knowing any thing of *heat*. The first dawn of intelligence consists in changes of feeling by which *differences* begin to be recognized. Mind commences in this perception of differences; it cannot be said that we know any thing *of itself*, but only the differences between it and other things. And, as in bodily growth, so in mental growth, there is an assimilation of like with like, or a process of *integration*. From the very first, along with the perception of difference, there has been also a perception of likeness. The clock-stroke, when first heard, is felt simply as an impression *differing* from others that precede and succeed it in the consciousness; but, when heard again, not only is there this recognition of difference, but it is perceived as *like* the clock-stroke which preceded it. This second impression is assimilated to the first, and, when a third arises, it also coalesces with the former like impressions. And so of all other sights, sounds, and touches. Under the influence of constant changes of impression, and a constant assimilation of like with like, there arise, at first vague, and then distinct unlikenesses among the feelings; that is, sights begin to be distinguished from sounds, and sounds from touches, while, at the same time, differences begin to be perceived among the impressions of each sense. In this way, the consciousness, at first homogeneous, grows into diversity, or becomes more *heterogeneous*, while its separated or differentiated parts are termed *ideas*.

Let us look into this a little more closely. When an infant opens its eyes for the first time upon the flame of a candle, for example, an image is formed, an impression produced, and there is a change of feeling. But the flame is not known, because there is as yet no *idea*. The trace left by the first impression is so faint that, when the light is removed, it is not remembered; that is, it has not yet become a mental possession. As the light, however, flashes into its eyes a great many times in a few weeks, each new impression is added to the trace of former impressions left in the nervous matter, and thus the impression deepens, until it becomes so strong as to remain when the candle is withdrawn. The idea therefore grows by exactly the same process as a bone grows; that is, by the successive incorporation of like with like. By the integration of a long series of similar impressions, one portion of consciousness thus becomes differentiated from the rest, and there emerges the *idea* of the flame. Time and repetition are therefore the indispensable conditions of the process.*

Now, when the candle is brought, the child recognizes or knows it; that is, it perceives it to be *like* the whole series of impressions of the candle-flame formerly experienced. It knows it because the impression produced agrees with the idea. In this way, by numerous repetitions of impressions, the child's first ideas arise; and, in this way, all objects are known. We know things, because, when we see, hear, touch, or taste them, the present impression spontaneously blends with like impressions before experienced. We know or recognize an external object not by the single impression it produces, but because

---

* "The single taste of sugar, by repetition, impresses the mind more and more, and, by this circumstance, becomes gradually easier to retain in idea. The smell of a rose, in like manner, after a thousand repetitions, comes much nearer to an independent ideal persistence than after twenty repetitions. So it is with all the senses, high and low. Apart altogether from the association of two or more distinct sensations, in a group or in a train, there is a fixing process going on with every individual sensation, rendering it more easy to retain when the original has passed away, and more vivid when by means of association it is afterward reproduced. This is one great part of the education of the senses. The simplest impression that can be made of taste, smell, touch, hearing, sight, needs repetition in order to endure of its own accord; even in the most persistent sense—the sense of seeing—the impressions on the infant mind that do not stir a strong feeling will vanish as soon as the eye is turned some other way."—*Professor Bain.*

that impression revives a whole train or group of previous discriminations that are like or related to it; while the number of those that are called up is the measure of our intelligence regarding it. If something is seen, heard, felt, or tasted, which links itself to no kindred idea, we say "we do not know it;" if it partially agrees with an idea, or revives a few discriminations, we know something about it, and the completer the agreement the more perfect the knowledge.

As to know a thing is to perceive its differences *from* other things, and its likeness *to* other things, it is therefore strictly an act of *classing*. This is involved in every act of thought, for to recognize a thing is to classify its impression or idea with previous states of feeling. Classification, in all its aspects and applications, is but the putting together of things that are alike—the grouping of objects by their resemblances; and as to know a thing is to know that it is *this* or *that*, to know what it is like and what it is unlike, we begin to classify as soon as we begin to think. When the child learns to know a tree, for example, he discriminates it from objects that differ from it, and identifies it with those that resemble it; and this is simply to class it as a tree. When he becomes more intelligent regarding it—when, for instance, he sees that it is an elm or an apple-tree—he simply perceives a larger number of characters of likeness and difference.

How our *degrees* of knowledge resolve themselves into successive classifications has been well illustrated by Herbert Spencer. He says: "The same object may, according as the distance or the degree of light permits, be identified as a particular negro; or, more generally, as a negro; or, more generally still, as a man; or, yet more generally, as some living creature; or most generally, as a solid body; in each of which cases the implication is, that the present impression is like a certain order of past impressions."

In early infancy, when the mind is first making the acquaintance of outward things, mental growth consists essentially in the production of *new ideas* by repetition of sensations, although such ideas never arise singly, but are always linked together in their origin. But, when a stock of ideas has been formed in this manner, the mental growth is mainly carried forward by new *combinations* among them. The simpler ideas once acquired, the development of intelligence consists largely in associating them in new relations and groups of relations. The perception of likeness and difference is the essential work that is going on all the time, but the comparisons and discriminations are constantly becoming more extensive, more minute, and more accurate. A number of elementary ideas thus become, as it were, fused or consolidated into one complex idea; and, by a still further recognition of likeness and difference, this is classed with a new group, and this again with still larger clusters of associated ideas.

The conception of an orange, for example, is compounded of the elementary notions of color, form, size, roughness, resistance, weight, odor, and taste. These elements are all bound up in one complex idea. The idea of an apple, a pear, a peach, or a plum, is in each case made up of a different group of component ideas, while the notion of a basket of different fruits is a cluster of these groups of still higher complexity, but still represented in thought as one complex idea, the elements of which are united by the relations of contrast and resemblance. Or, again, the child may begin with a large, vague idea, as a tree, for example, and then, as intelligence concerning it progresses, he decomposes it into its component ideas, as trunk, branches, leaves, roots, and these into still minuter parts. There is a growing mental heterogeneity through the increasing perception of likeness and difference. Thus, as soon as ideas are formed, they begin to be used over and over, and this process is ever continued.* An old idea in a new relation or grouping has a new meaning—becomes a new fact or

---

* Our reason consists in using an old fact in new circumstances, through the power of discerning the agreement; this is a vast saving of the labor of acquisition; a reduction of the number of original growths requisite for our education. When we have any thing new to learn, as a new piece of music, or a new proposition in Euclid, we fall back upon our previously-formed combinations, musical or geometrical, so far as they will apply, and merely tack certain of them together in correspondence with the new case. The method of acquiring by patch-work sets in early, and predominates increasingly.— *Bain.*

a new truth. The perception of new resemblances and of new differences gives rise to new groupings and new classings of ideas, and thus the mind grows into a complex and highly-differentiated organism of intelligence, in which the internal order of thought-relations answers to the external order of relations among things.

That which occurs at this earliest stage of mental growth is exactly what takes place in the *whole course* of unfolding intelligence. Simple as these operations may seem, and begun by the infant as soon as it is born, in their growing complexities, they constitute the whole fabric of the intellect. What we term the "mental faculties" are not the ultimate elements of mind, but only different modes of the mental activity; and, as one law of growth evolves all the various organs and tissues of the bodily structure, so one law of growth evolves all the diversified "faculties" of the mental structure. Under psychological analysis, the operations of reason, judgment, imagination, calculation, and the acquisitions of the most advanced minds yield at last the same simple elements—the perceptions of likenesses and differences among things thought about; while memory is simply the power of *reviving* these distinctions in consciousness. Whatever the object of thought, to know in what respects it differs from all other things, and in what respects it resembles them, is to know all about it—is to exhaust the action of the intellect upon it. The way the child gets its early knowledge is the way *all* real knowledge is obtained. When it discovers the likeness between sugar, cake, and certain fruits, that is, when it integrates them in thought as *sweet*, it is making just such an induction as Newton made in discovering the law of gravitation, which was but to discover the likeness among celestial and terrestrial motions. And as with physical objects, so also with human actions. The child may run around the house and play with its toys; it must not break things or play with the fire. Here, again, are relations of likeness and unlikeness, forming a basis of moral classification. The judge on the bench is constantly doing the same thing; that is, tracing out the likenesses of given actions, and classing them as right or wrong.*

Having thus formed some idea of how mental growth takes place, let us now roughly note how far it proceeds in the first three or four years of childhood.

III.—EXTENT OF EARLY MENTAL GROWTH.

From the hour of birth, through all the waking moments, there pour in through the eye ever-varying impressions of light and color, from the dimness of twilight to the utmost solar refulgence, which are reproduced as a highly-diversified luminous consciousness. Impressions of sound of all qualities and intensities, loud and faint, shrill and dull, harsh and musical, in endless succession, enter the ear, and give rise to a varied auditory consciousness. Ever-changing contrasts of touch acquaint the mind with hard things and soft, light and heavy, rough and smooth, round, angular, brittle, and flexible, and are wrought into a knowledge of things within reach. And so, also, with the senses of taste and smell. This multitude of contrasted impressions, representing the endless diversity of the surrounding world, has been organized into a connected and coherent body of knowledge.

After two or three years the face that was at first blank becomes bright with the light of numberless recognitions. The child knows all the common objects of the house, the garden, and the street, and it not only knows them apart, but it has extended its discriminations of likeness and difference to a great many of their characters. It has found out about differences and resemblances of form, size, color, weight, transparency, plasticity, toughness, brittleness, fluidity, warmth, taste, and various other properties of the solid and liquid sub-

---

* To those who care to pursue this important subject of mental growth, which is here hardly more than hinted at, I would recommend the "Principles of Psychology," by Mr. Herbert Spencer, now being published in parts by D. Appleton & Co. Mr. Spencer considers mind from the point of view of *Evolution*, and his work is, beyond doubt, the most important contribution to this aspect of psychological science that has yet been made. I have to acknowledge my own indebtedness to it.

stances of which it has had experience. It has noted peculiarities among many animals and plants, and the distinctions, traits, and habits of persons.

Besides this, it has learned to associate names with its ideas; it has acquired a language. The number of words it uses to express things and actions, and qualities, degrees, and relations, among these things and actions, shows the extent to which its discriminations have been carried. Groups of ideas are integrated into trains of thought, and words into corresponding trains of sentences to communicate them. Nor is this all. There is still another order of acquisitions in which the child has made remarkable proficiency. The infant is endowed with a spontaneous activity: it moves, struggles, and throws about its limbs as soon as it is born. But its actions are at first aimless and confused. As it knows nothing, of course, it can *do* nothing; but, with the growth of distinct ideas and feelings, there is also a growth of special movements in connection with them. It has to find out by innumerable trials how to creep, to walk, to hold things, and to feed itself. To see an object and to be able to seize it, or to go and get it, result from an adjustment of visual impressions with muscular movements, which it has taken thousands of experiments to bring under control. The vocal apparatus has been brought under such marvellous command that hundreds of different words are uttered, each requiring a different combination of movements of the chest, larynx, tongue, and lips. Numerous aptitudes and dexterities are achieved, and, when, stimulated by curiosity, it examines its toy and breaks it open to find "what makes it go," it has entered upon a career of active experiment, as truly as the man of science in his laboratory.

IV.—NATURE'S EDUCATIONAL METHOD.

Such is Nature's method of education, and such its earliest results. Human beings are born into a world of stubborn realities; of laws that are fraught with life and death in their inflexible course. What the new-born creature shall be taught is too important to be left to any contingency, and so

NATURE'S EDUCATIONAL METHOD.

Nature takes in hand the early training of the whole human race, and secures that rudimentary knowledge of the properties of things which is alike indispensable to all. It is, however, only the obvious characters and simpler relations of objects which are thrust conspicuously upon the attention that are recognized in childhood. But the method of bringing out mind has been established. Nature's early tuition has given shape to the mental constitution, and determined the conditions and order of its future development. The child is sent to school—the school of experience—as soon as it is born, and Nature's method of leading out the intelligence is that of *growth*. She roots mental activity in organic processes, and thus *times* the rate of acquisition to the march of organic changes. She is never in haste, but always at work; never crams, but ever repeats, assimilates, and organizes. Her policy of producing vast effects by simple means is not departed from in the realm of mind; indeed, it is more marvellous here than anywhere else. While the organic world is made up almost entirely of but four chemical elements, the intellectual world is constituted wholly of but *two* ultimate elements, the perception of likeness and the perception of difference among objects of thought. These elements are wrought into the mental constitution through the direct observation and experience of things. Mind is called forth by the spontaneous interaction of the growing organism and the agencies and objects of surrounding Nature.

The school-period at length arrives, and Art comes forward to assume the direction of processes that Nature has thus far conducted. But her course is plainly mapped out; the work begun is to be continued. New helps and resources may be needed, but the end and the essential means should be the same. Mental growth is to be carried by cultivation to still higher stages, but by the same processes hitherto employed. The discriminations of likeness and difference by which all things are known, the comparison, classification, and association of ideas in which knowledge arises, are to become more accurate, more extensive, and more systematic. To do this the mind is to be maintained in living contact with the realities which environ it, but which are now to be regularly studied. We have here the clear criterion by which educational systems must be judged; how does the prevailing practice answer to the test?

V.—DEFICIENCY OF EXISTING SCHOOL-METHODS.

Nothing is more obvious than that the child's entrance upon school-life, instead of being the wise continuation of processes already begun, is usually an abrupt transition to a new, artificial, and totally different sphere of mental experience. Although, in the previous period, it has learned more than it ever will again in the same time, and learned it according to the fundamental laws of growing intelligence, yet the current notion is, that education *begins* with the child's entrance upon school-life. How erroneous this is we have sufficiently seen. That which does begin at this time is not *education*, but simply the acquirement of new helps to it. The first thing at school is usually the study of words, spelling, reading, and writing—that is, to get the use of written language. This is, of course, important and indispensable. To be able to accumulate, compare, arrange, and preserve ideas, and put them to their largest uses, it is necessary to *mark* them. Words are these marks or signs of ideas, and, as such, have an inestimable value. Words, as the marks of ideas, are the representatives of knowledge, and books which contain them become the invaluable depositories of the world's accumulating thought. It is exactly because of their great importance and their intimate relations to our intellectual life, that we should be always vividly conscious of their exact nature and office.

But words are not ideas, they are only the *symbols* of ideas; language is not knowledge, but the *representative* of it. Labels have a value of convenience, which depends upon the intrinsic value of what they point out. Now, there is a constant and insidious tendency in education to invert these relations—to exalt the husk above its contents, the tools above their work, the label above its object, words above the things for which they stand. The *means* of culture thus become the *ends* of culture, and education is emptied of its substantial purpose. In the lower institutions, while that acquisition and organization of ideas in which education really consists are neglected, to spell accurately, to read fluently, to define promptly, and to write neatly, are the ideals of school-room accomplishment. In the higher institutions, this ideal expands into the proficient command of a multitude of words, and skill in the arts of expression, so that the student piles language upon language until he has tagged half a dozen labels to each of his scanty, and ill-conceived ideas.

The glaring deficiency of our popular systems of instruction is, that words are not subordinated to their real purposes, but are permitted to usurp that supreme attention which should be given to the formation of ideas by the study of things. It is at this point that true mental growth is checked, and the minds of children are switched off from the main line of natural development into a course of artificial acquisition, in which the semblance of knowledge takes the place of the reality of knowledge.

We have seen that the growth of mind results from the exercise of its powers upon the direct objects of experience, and consists in its recognition of distinctions among the properties and relations of things, and in the classing and organization of ideas thus acquired. These operations can be facilitated by the use of words and books, but only when the ideas themselves are first clearly conceived as the accurate representations of things. But the ordinary word-studies of our schools, which are truly designed to *assist* these operations, are actually made to *exclude* them. The child glides into the habit of accepting words *for* ideas, and thus evades those mental actions which are only to be performed upon the ideas themselves.

The existing systems of instruction are therefore deficient, by making no adequate provision for cultivating the growth of ideas by the exercise of the observing powers of children. Observation, the capacity of recognizing distinctions, and of being mentally alive to the objects and actions around us, is only to be acquired by practice, and therefore requires to become a regular and habitual mental exercise, and to have a fundamental place in education.

The importance of training the young mind to habits of correct observation, to form judgments of things noted, and to describe correctly the results of observation, can hardly be over-estimated. It has been well remarked that, "without an accurate acquaintance with the visible and tangible properties of things, our conceptions must be erroneous, our inferences fallacious, and our operations unsuccessful. The education of the senses neglected, all after-education partakes of a drowsiness, a haziness, an insufficiency, which it is impossible to cure. Indeed, if we consider it, we shall find that exhaustive observation is an element of all great success. It is not to artists, naturalists, and men of science only, that it is needful; it is not only that the skilful physician depends on it for the correctness of his diagnosis, and that to the good engineer it is so important, that some years in the workshop are prescribed for him; but we may see that the philosopher also is fundamentally one who *observes* relationships of things which others had overlooked, and that the poet, too, is one who *sees* the fine facts in Nature which all recognize when pointed out, but did not before remark. Nothing requires more to be insisted on than that vivid and complete impressions are all-essential. No sound fabric of wisdom can be woven out of a rotten, raw material."

It needs hardly to be repeated that observation is the starting-point of knowledge, and the basis of judgment and inductive reasoning. In the chaos of opinions among men, the conflicts are usually on the *data*, which have not been observed with sufficient care. Dispute is endless until the facts are known, and, when this happens, dispute is generally ended. Dr. Cullen, long ago, remarked: "There are more *false facts* in the world than false hypotheses to explain them; there is, in truth, nothing that men seem to admit so lightly as an asserted fact."

Children should, therefore, be taught to *see for themselves*, and to think for themselves on the basis of what they have seen. In this way only can they learn to weigh the true value of evidence, and to guard against that carelessness of assumption and that credulous confidence in the loose statements of others, which is one of the gross mental deficiencies we everywhere encounter. This is one of the rights of the understanding too little respected in the school-room. Instead of being called into independent activity, children's minds are rather repressed by authority. In the whole system of word-teaching the statements have to be taken on trust. "This is the rule," and "that the usage," and the say-so of book and teacher is final. Granted that much, at any rate, in education is to be accepted on authority, it is all the more necessary that there should be, in some departments, such an assiduous cultivation of personal observation and independent judgment as may serve to guard against errors from this source.

It may be said that arithmetic forms an exception to what is here stated respecting the prevalence of authority in schools, as its operations are capable of independent proof. This is true, but the exception is of such a nature that it cannot serve as a *correction;* for it reasons not from observed facts, but from assumed numerical data. Mathematics, says Prof. Huxley, " is that study which knows nothing of observation, nothing of induction, nothing of experiment, nothing of causation."

The foregoing strictures, I am aware, have a variable applicability to different schools. Many teachers are alive to these evils, and strive in various ways to mitigate them; but the statement, nevertheless, holds sadly true in its general application. There is a radical deficiency of existing educational methods which cannot be supplied by the mere make-shift ingenuity of instructors, but requires some systematic and effectual measure of relief.

VI.—WHAT IS NOW MOST NEEDED.

To supply this unquestionable deficiency, we should demand the introduction into primary education, in addition to reading, writing, and arithmetic, of A FOURTH FUNDAMENTAL BRANCH OF STUDY, WHICH SHALL AFFORD A SYSTEMATIC TRAINING OF THE OBSERVING POWERS. We are entitled to require that, when the child enters school, it shall not take leave of the universe of fact and law, but that its mind shall be kept in intimate relation with Nature in some one of her great divisions, and that the knowledge acquired shall be actual and thorough, and suited to call out those operations which are essential to higher mental growth. It is agreed by many of the ablest thinkers that such an element of mental training is now the urgent want of general education. Dr. Whewell thus defines the present need:

" One obvious mode of effecting this discipline of the mind is the exact and solid study of some portion of inductive knowledge. . . . botany, comparative anatomy, geology, chemistry, for instance. But I say, the *exact* and *solid* knowledge; not a mere verbal knowledge, but a knowledge which is real in its character, though it may be elementary and limited in its extent. The knowledge of which I speak must be a knowledge of things, and not merely of names of things; an acquaintance with the operations and productions of Nature as they appear to the eye; not merely an acquaintance with what has been said about them; a knowledge of the laws of Nature, seen in special experiments and observations before they are conceived in general terms; a knowledge of the types of natural forms, gathered from individual cases already familiar. By such study of one or more departments of inductive knowledge, the mind may escape from the thraldom and illusion which reigns in the world of mere words."

The increasing influence of science over the course of the world's affairs is undeniable. Not only has it already become a controlling force in civilization, but it is steadily invading the higher spheres of thought, and, by its constant revisions and extensions of knowledge, it is rapidly reshaping the opinion of the world. That such an agency is destined to exert a powerful influence upon the culture of the human mind, is inevitable. Already, indeed, it has become a recognized element of general instruction, but it has been pursued in such a fragmentary and incoherent way, that its legitimate mental influence is far from having been realized. The immediate problem, then, is how to organize the scientific element of study so as to gain its benefits, as a mental discipline. Each of the prominent sciences—physics, chemistry, geology, botany—has its special advantages, and is entitled to a place in a liberal course of study. But some one must be selected which is best fitted to be generally introduced into primary schools. The work must begin here, if it is to be thoroughly done.

The system of teaching by object-lessons is an attempt to meet the present requirement in the sphere of primary education. But these efforts have been rather well-intentioned gropings after a desirable result than satisfactory realizations of it. The method is theoretically correct, and some benefit cannot fail to have resulted; but the practice has proved incoherent, desultory, and totally insufficient as a *training* of the observing powers. Nor can this be otherwise so long as all sorts of objects are made to serve as "lessons," while the exercises consist merely in learning a few obvious and unrelated characters. Although, in infancy, objects are presented at random, yet, if mental growth is to be definitely directed, they must be presented in relation. A lesson one day on a bone, the next on a piece of lead, and the next on a flower, may be excellent for imparting "information," but the lack of relation among these objects unfits them to be employed for developing connected and dependent thought. This teaching can be thoroughly successful only where the "objects" studied are connected together in a large, complex whole, as a part of the order of Nature. The elementary details must be such as children can readily apprehend, while the characters and relations are so varied and numerous as to permit an extended course of acquisition issuing in a large body of scientific principles. Only in a field so broad and inexhaustible as to give play to the mental activities in their continuous expansion can object-studies have that real disciplinary influence which is now so desirable an element of popular education.

What we most urgently need is an objective course of study which shall train the observing powers *as mathematics trains the power of calculation.* From the time the child begins to count, until the man has mastered the calculus, there is provided an unbroken series of exercises of ever-increasing complexity, suited to unfold the mathematical faculty. We want a parallel course of objective exercises, not to be dispatched in a term or a year, but running through the whole period of education, which shall give the observing and inductive faculties a corresponding continuous and systematic unfolding. What subject is best fitted for this purpose?

VII.—ADVANTAGES OFFERED BY BOTANY.

The largest number of advantages for the purpose we have in view will be found combined in that branch of natural history which treats of the vegetable kingdom. While each of the sciences has its special claim as a subject of study, it is thought that none of them can compare with Botany in fulfilling the various conditions now indicated, and which entitle it to take a regular and fundamental place in our scheme of common-school instruction. Its prominent claims are:

I. The materials furnished by the vegetable kingdom for direct observation and practical study are abundant, and easily accessible, overhead, underfoot, and all around—grass, weeds, flowers, trees—open and common to everybody. There is no expense, as in experimental science. And, in meeting this fundamental condition of a universal objective study, it may be claimed that Botany is without a rival.

II. The collection of specimens may be carried on as regularly as any other school-exercise, while they are just as suitable objects upon the scholar's desk as the books themselves. They cannot interfere with the order and propriety of the class-room.

III. The elementary facts of Botany are so simple, that their study can be commenced in early childhood, and so numerous as to sustain a prolonged course of observation. The characters of plants which engage attention at this period of acquisition are external, requiring neither magnifying-glass nor dissecting-knife to find them.

IV. From these rudimentary facts the pupil may proceed gradually to the more complex, from the concrete to the abstract—from observations to the truths that rest upon observation, in a natural order of ascent, as required by the laws of mental growth. If properly commenced, the study may be stopped at any stage, and the advantages gained are substantial and valuable, while, at the same time, it is capable of tasking the highest intelligence through a lifetime of study.

V. The means are thus furnished for organizing object-teaching into a systematic method, so that it may be pursued definitely and constantly through a course of successively higher and more comprehensive exercises.

VI. Botany is unrivalled in the scope it offers to the cultivation of the descriptive powers, as its vocabulary is more copious, precise, and well-settled than that of any other of the natural sciences. Upon this point—most important in its educational aspect—Prof. Arthur Henfrey has well remarked:

"The technical language of Botany, as elaborated by Linnæus and his school, has long been the admiration of logical and philosophical writers, and has been carried to great perfection. Every word has its definition, and can convey one notion to those who have once mastered the language. The technicalities, therefore, of botanical language, which are vulgarly regarded as imperfections, and as repulsive to the inquirer, are, in reality, the very marks of its completeness, and, far from offering a reason for withholding the science from ordinary education, constitute its great recommendation as a method of training in accuracy of expression and habits of describing definitely and unequivocally the observations made by the senses. The acquisition of the terms applied to the different parts of plants exercises the memory, while the mastery of the use of the adjectives of terminology cultivates, in a most beneficial manner, a habit of accuracy and perspicuity in the use of language."

Botanical language is the most perfect that is applied to the description of external nature, but its accuracy is not the accuracy of geometry, the terms of which call up the same sharply-defined invariable conceptions. But the characters of natural objects are not such rigid and exact repetitions of each other. Nature is constantly varying her types. The application of botanical terms is, therefore, not a mere mechanical act of the mind, but involves the exercise of *judgment*.

VII. It is congenial with the pleasurable activity of childhood, and makes that activity subservient to mental ends. It enforces rambles and excursions in quest of specimens, and thus tends to relieve the sedentary confinement of the school-room, and to promote health by moderate open-air exercise.

VIII. The knowledge it imparts has a practical value in various important directions. It is indispensable to the intelligent pursuit of agriculture and horticulture—avocations in which more people are occupied and interested than in all others put together.

IX. The study of plant-forms opens to us a world of grace, harmony, and beauty, that is not without influence upon the æsthetic feelings, and the appreciation of art. Intimately involved as is the vegetable kingdom with the ever-changing aspects of Nature, it is well fitted to attract the mind to the fine features of scenery, and the grand effects of the natural world.

X. Knowledge of this subject is a source of pure and unfailing personal enjoyment. Its objects constantly invite attention, and vary more or less with each locality, so that the botanical student is always at home, and is always solicited by something fresh and attractive.

XI. The pursuit of Botany to its finer facts and subtler revelations involves the mastery of the microscope—one of the most delicate and powerful of all instruments of observation. It also opens the field of experiment, and affords opportunity for cultivating manipulatory processes.

XII. Notwithstanding the superficial prejudice against Botany, as a kind of light, fancy subject, dealing with flowers—an "accomplishment" of girls—it is nevertheless a solid and noble branch of knowledge. It has intimate connections with all the other sciences—physics, chemistry, geology, meterology, and physical geography—helps them all, and is helped by all. It treats of the phenomena of organization, and is the proper introduction to the great subject of Biology—the science of the general laws of life.

These considerations show that, for the purpose we have in view—the introduction of a subject into education which shall extend through all its grades, and afford a methodical discipline in the study of things—Botany has eminent, if not unrivalled claims to the attention of educators.

VIII.—DEFECTS OF COMMON BOTANICAL STUDY.

But the benefits here sought are not to be gained by the usual way of dealing with the subject. For this end it must be pursued by the direct study of its objects, and in a definite order. The concrete and elementary character of plants *must* be made familiar before the truths based upon them can become real mental possessions. The common method of acquiring Botany, *in its results*, that is, by going at once to its general principles, is hence peculiarly futile for purposes of education. The mere reading up of vegetable physiology is no better than getting any other second-hand information. To learn a number of hard botanical terms without really knowing what they represent, or to con over classifications that are equally void of significance, is much the same as any other verbal cramming. The objection to ordinary botanical study is, not that the books do not tell the pupil a great many interesting and useful things about plants, but that he studies it as he does ancient history, treating its objects as if they had all gone to dust thousands of years ago.

Besides, that which goes under the *name* in many of our schools is not *Botany* in any true sense; it is only a *branch* of it. In the early part of the century, the subject had become so overgrown with the mere pedantries of naming, that there came a reaction against systematic Botany, or the study of the relationships of plants, and some went so far as to insist that the whole science could be "evolved" by studying a single plant. Under the influence of this tendency, Botany became merged in the study of vegetable physiology to the neglect of its descriptive and relational elements. But it is now recognized that all parts of the science are intimately correlated, and that the inner relations of plants can only be well understood by first getting a knowledge of their outer relations. Nevertheless, the tendency to sink it in mere physiology was strongly felt in education, which instinctively seized upon a view of the subject most easily got through books. But vegetable physiology is not Botany any more than the rule of three is arithmetic; and to engage with the body of generalized truths, which make up the higher parts of the science, before first mastering Descriptive Botany, is like attacking the higher problems of arithmetic before learning its simple rules.

Nor is the case much helped by that casual inspection of specimens in which students sometimes indulge. To pick a flower to pieces now and then, or to identify a few plants by the aid of glossaries and analytical tables, and to press and label them, are, no doubt, useful operations, but they are far from answering the educational purposes here contemplated.

IX.—AIMS OF THE PRESENT WORK.

In the preparation of the present work, the end kept strictly in view has been to make it conform to the laws of mental growth. Although it attempts to make a beginning only, yet it claims to begin right—to teach Botany as it should be taught, and, in so doing, to cultivate systematically those parts of the mind which general education most neglects. It is adapted to these purposes in the following respects:

In the first place it conforms to the method of Nature by making actual phenomena the objects of thought. It continues the direct intercourse of the mind with things, by selecting that portion of the natural world which seems best adapted for the purpose, and providing for its direct and regular study. It is a merit of the plan that it permits no evasion of this purpose, but *compels* attention to the objects selected. There are no lessons to "commit and recite;" the child's proper work being to observe, distinguish, compare, and describe; and thus, from the outset, it is exercising its own faculties in the organization of real knowledge.

In the second place, the present plan implies that habits of regular observation shall be commenced *early*. This is on various accounts a most important feature. The child should begin to be taught *how* to notice, and *what* to look for, because it is already spontaneously engaged in the work, and needs guidance. While its mental life is (so to speak) external, and it hungers for changing impressions and new sensations, is certainly the time to foster and direct this activity. It is necessary to furnish abundant and varied materials for simple observation in this impressible sensational stage of mental growth, when, as yet, only rudimentary details can be appreciated. At this time they can be rapidly acquired and easily remembered, while, as the mind advances to the reflective stage, unless the habit of observation has been formed, attention to details becomes tedious and irksome.

It is sometimes said that it is absurd to attempt teaching children "science" before twelve or fourteen years of age; and, if it be meant the memorizing of the principles and results of science, the remark is true. But it is not true if applied to the early observation of those simple facts which lead up to scientific principles. Nature settles all that by putting children to the study of the properties of natural objects as soon as they are born. The germ of science is involved in its earliest discriminations. When the child first distinguishes its father from its mother, it is doing the same thing that Leverrier did in distinguishing Neptune from a fixed star; the difference is only one of *degree*. In putting children early to the work of observation, as is provided for in this little work, we are, therefore, only continuing a course already entered upon, and which involves the most natural and congenial action of the childish mind.

Another reason why children should commence the study of objects early is, that the *habit* may be formed before the mind acquires a bent in other directions; is, because to postpone it is to defeat it. As education is supposed to begin when school begins, and to consist mainly in learning lessons, children quickly get the notion that nothing is properly "education" that does not come from books. But the difficulty here is deeper still. The habit of lesson-learning, of passively loading the memory with verbal acquisitions, is so totally different a form of mental action from observing, inquiring, finding things out, and judging independently about them, that the former method tends powerfully to hinder and exclude the latter. I have found, in my own experience, that the younger children took to exercises in observation with freedom, and zest, while their elders, in proportion to their school proficiency, had to overcome something of both disinclination and disqualification for the work.

In the third place, the plan of study here proposed recognizes the importance in education of the element of *time*. The very conception of mental unfolding as a *growth* implies, as we have seen, an orderly succession of natural processes to which *time* is an indispensable condition. Ideas are not only to be obtained by observation, but they are to be organized into knowledge. That this may be done effectually, so that acquisitions shall be lasting, it must be done slowly and by numberless repetitions. The plan of this First Book complies with this condition by such a construction of the exercises as will secure constant repetition and a thorough assimilation of observations.

It complies with the time-requirement in another respect also: it is but a *first* step, and involves many succeeding steps. The mind grows, let it be remembered, for twenty or thirty years, passing through successive phases, in which now one form of mental action predominates, and now another. Every study, which aims to cultivate any class of mental activities up to the point of *discipline*, must extend through a considerable part of this period. This is well understood with respect to mathematics and Latin; they run through from the ages of seven or eight years to college graduation; while *three months* is the usual collegiate allowance of time for Botany. As the true mode of treating the subject, both on its own account and for educational purposes, requires that it be pursued in a definite order through the whole school career, I have here conformed to that condition by presenting only the first rudimentary instalment of the subject.

Fourthly and finally, the mode of study here proposed is specially suited to call forth those operations in which growing intelligence consists.

A child old enough to begin the study of Botany has already acquired a large stock of ideas of concrete things and their relations. As concerns plants, it has probably discriminated between leaves, flowers, stems, and roots. Its idea of a leaf, for instance, though loose and indefinite, is still roughly correct. The thin, green plate contrasts strongly with the other parts of the plant. Its differences from flowers and stems enable the mind readily to differentiate it in idea, while the essential resemblances of leaves of all kinds make their integration into one general conception inevitable.

Our primary scholar, then, starting at the level of ordinary perception, is to increase his discriminative power. He must learn to discover minuter differences and resemblances, and to make his ideas more definite and precise. To this end he enters upon the first exercises of this work, and begins a careful inspection of leaves. He soon finds that they vary considerably; that their most conspicuous feature—that which he has hitherto regarded as the *entire leaf*—forms, in most cases, but *one part* of the leaf. Having gained a clear idea of this part, he marks his conception of it by a sign which he finds to be the word *blade*. Another part, almost always present, he distinguishes as the leaf-stem, and names it the *petiole;* and still another part, probably never before noticed, he learns to recognize as the *stipules*.

He thus begins with the recognition of simple differences, the idea of the leaf being resolved into three component ideas. But each of these component ideas is crude from lack of observation of the varying forms of different blades, petioles, and stipules. Observation is now extended to new specimens, and as it goes forward new differences are perceived among these parts. The blade turns out to be composed of different elements. Its framework is differentiated from its soft, pulpy covering, receives its name, and then *this part* opens a new field of observation in recognizing and comparing the different modes and variations of the *venation*, as it is called.

In this way there grows up an intelligent conception of the leaf. Its idea, at first vague and homogeneous, by successive discriminations of differences and resemblances has become definite and heterogeneous. The conception, at first simple, is now complex, but it is an orderly complexity, in which the parts of the object, with the relations of those parts, are distinctly possessed in thought. After a month of observation so conducted, in which numerous specimens are observed and compared, and their peculiarities noted and named, the pupil will have begun to acquire some facility in observing and describing, and will have gained a good deal of knowledge of this elementary portion of the subject.

Having gone over simple and compound leaves, he next passes to the examination of the stem. Here, also, his first notion is simple and indefinite, but, when a good many have been noticed, marked differences of appearance present themselves, and stems begin to fall into groups, which he describes as round, square, erect, trailing, creeping, etc., as the case may be; while closer observation reveals still minuter characters of difference and resemblance among them. Inflorescence, flowers, and roots, are successively studied in the same manner.

Beginning thus with the rudimentary characters of the simplest parts, the child proceeds step by step, until he becomes acquainted with the leading characters of the plant as a whole, while the faculties drawn out, and the work of drawing them out, conform to the first conditions of unfolding intelligence. A multitude of accurate botanical ideas have been obtained of the endless diversities of feature and form in the vegetable world, but they do not lie as a burden of details in the memory; they have been arranged into organized knowledge. Particular facts are gradually fused into general conceptions; each new peculiarity observed is a discrimination of difference or likeness which links itself to previous conception. The simplest ideas are at first associated in minor groups, and soon reappear in larger groupings and relations, until at length the whole plant has been reproduced in the mind as a highly complex organism of thought-relations—the mental representation being as truly a product of growth as the living object itself.

It has been explained that the first and simplest thinking involves the rudimentary act of classing. Botanical study, pursued in the direct way here practised, is specially fitted to cultivate this form of mental exercise, as Botany is eminently a classificatory science. Beginning with the simplest discriminations and comparisons, the pupil has arranged the characters observed into groups in accordance with their resemblances. As he becomes able to grasp in thought these assemblages of characters, and to discern remoter relations of likeness, the classification is carried further, and he is thus gradually prepared to go on and trace out those larger and more complex relationships of difference and resemblance by comparison of *all* the characters of plants, which lead to the complete classification of the vegetable kingdom on the natural system.

This mode of mental acquisition has also enforced a salutary training in the use of language. Words are used with more clearness and reality of meaning, and, instead of rehearsing what others have observed, he learns to describe what he has seen and knows himself. In this way he goes *behind* the words to the ideas, and things they symbolize, and can better appreciate both their value and their imperfection as signs. For example, upon first noting a plant character, he confidently applies a term to it; but, upon looking further, he perhaps fails to find the exact repetition of it, and doubt may arise about its new application. He soon discerns that Nature's plan is not that of sharp lines of distinction, but that she rather *flows* from character to character in ceaseless continuity, and never exactly repeats herself. Words are therefore no longer to be accepted as the *absolute equivalents* of things; they cannot represent the delicate shadings and the infinite variety of nature and of thought; they are but imperfect signs, frequently liable to mislead, and therefore demanding judgment in their use.

This exercise of judgment, which is constantly required in estimating characters and in making plant-descriptions, is of incalculable advantage to the young. Although the child's warrant for his statement is, "I saw it," yet he quickly learns that the main thing, after all, is, how the thing seen *is to be regarded*. He is constantly called upon to make up his mind; he will have frequently to suspend his opinion, and sometimes, perhaps, to maintain it against his teacher. But this is just the kind of mental work that he will have to do in after-life, in forming his conclusions upon subjects of familiar observation and practical experience.

THE CORRELATION AND CONSERVATION

OF

FORCES.

A SERIES OF EXPOSITIONS BY GROVE, MAYER, HELMHOLTZ, FARADAY, LIEBIG, AND CARPENTER.

WITH

AN INTRODUCTION.

BY E. L. YOUMANS.

The work embraces:

I.—THE CORRELATION OF PHYSICAL FORCES. By W. R. GROVE. (The complete work.)

II.—CELESTIAL DYNAMICS. By DR. J. R. MAYER.

III.—THE INTERACTION OF FORCES. By PROF. HELMHOLTZ.

IV.—THE CONNECTION AND EQUIVALENCE OF FORCES. By PROF. LIEBIG.

V.—ON THE CONSERVATION OF FORCE. By DR. FARADAY.

VI.—ON THE CORRELATION OF PHYSICAL AND VITAL FORCES. By DR. CARPENTER.

D. APPLETON & CO.'S PUBLICATIONS.

THE PHYSIOLOGY
AND
PATHOLOGY OF THE MIND.

By HENRY MAUDSLEY, M. D., London.

1 volume, 8vo.   Cloth.   Price, $4.00.

CONTENTS:

Part I.—The Physiology of the Mind.

CHAPTER 1. On the Method of the Study of the Mind.
" 2. The Mind and the Nervous System.
" 3. The Spinal Cord, or Tertiary Nervous Centres; or, Nervous Centres of Reflex Action.
" 4. Secondary Nervous Centres; or Sensory Ganglia; Sensorium Commune.
" 5. Hemispherical Ganglia; Cortical Cells of the Cerebral Hemispheres: Ideational Nervous Centres; Primary Nervous Centres; Intellectorium Commune.
" 6. The Emotions.
" 7. Volition.
" 8. Motor Nervous Centres, or Motorium Commune and Actuation or Effection.
" 9. Memory and Imagination.

Part II.—The Pathology of the Mind.

CHAP. 1. On the Causes of Insanity.
" 2. On the Insanity of Early Life.
" 3. On the Varieties of Insanity.
CHAP. 4. On the Pathology of Insanity.
" 5. On the Diagnosis of Insanity.
" 6. On the Prognosis of Insanity.
CHAPTER 7. On the Treatment of Insanity.

"The first part of this work may be considered as embodying the most advanced expression of the new school in physiological psychology, which has arisen in Europe, and of which Bain, Spencer, Leycoch, and Carpenter, are the more eminent English representatives."—*Home Journal.*

"The author has professionally studied all the varieties of insanity, and the seven chapters he devotes to the subject are invaluable to the physician, and full of important suggestions to the metaphysician."—*Boston Transcript.*

"In the recital of the causes of insanity, as found in peculiarities of civilization, of religion, of age, sex, condition, and particularly in the engrossing pursuit of wealth, this calm, scientific work has the solemnity of a hundred sermons; and after going down into this exploration of the mysteries of our being, we shall come up into active life again chastened, thoughtful, and feeling, perhaps, as we never felt before, how fearfully and wonderfully we are made."—*Evening Gazette.*

D. APPLETON & CO.'S PUBLICATIONS.

THE PHYSIOLOGY OF MAN;

DESIGNED TO

REPRESENT THE EXISTING STATE OF PHYSIOLOGICAL SCIENCE AS APPLIED TO THE FUNCTIONS OF THE HUMAN BODY.

By AUSTIN FLINT, Jr., M. D.

Alimentation; Digestion; Absorption; Lymph and Chyle.

1 volume, 8vo. Cloth. Price, $4.50.

RECENTLY PUBLISHED.

THE FIRST VOLUME OF THE SERIES

BY

AUSTIN FLINT, Jr., M. D.,

CONTAINING

Introduction; The Blood; The Circulation; Respiration.

1 volume, 8vo. Cloth. Price, $4.50.

"Professor Flint is engaged in the preparation of an extended work on human physiology, in which he professes to consider all the subjects usually regarded as belonging to that department of physical science. The work will be divided into separate and distinct parts, but the several volumes in which it is to be published will form a connected series."—*Providence Journal.*

It is free from technicalities and purely professional terms, and instead of only being adapted to the use of the medical faculty, will be found of interest to the general reader who desires clear and concise information on the subject of man physical."—*Evening Post.*

"Digestion is too little understood, indigestion too extensively suffered, to render this a work of supererogation. Stomachs will have their revenge, sooner or later, if Nature's laws are infringed upon through ignorance or stubbornness, and it is well that all should understand how the penalty for 'high living' is assessed."—*Chicago Evening Journal.*

"A year has elapsed since Dr. Flint published the first part of his great work upon human physiology. It was an admirable treatise —distinct in itself—exhausting the special subjects upon which it treated."—*Philadelphia Inquirer.*

Works of *Herbert Spencer* published by *D. Appleton & Co.*

The Philosophy of Herbert Spencer.

FIRST PRINCIPLES;

IN TWO PARTS:

I. THE UNKNOWABLE. II. LAWS OF THE KNOWABLE.

In one Volume. 518 pages.

"Mr. Spencer has earned an eminent and commanding position as a metaphysician, and his ability, earnestness, and profundity, are in none of his former volumes so conspicuous as in this. There is not a crude thought, a flippant fling, or an irreverent insinuation in this book, notwithstanding that it has something of the character of a daring and determined raid upon the old philosophies."—*Chicago Journal.*

"This volume, treating of First Principles, like all Mr. Spencer's writings that have fallen under our observation, is distinguished for clearness, earnestness, candor, and that originality and fearlessness which ever mark the true philosophical spirit. His treatment of theological opinions is reverent and respectful, and his suggestions and arguments are such as to deserve, as they will compel, the earnest attention of all thoughtful students of first truths. Agreeing with Hamilton and Mansel in the general, on the unknowableness of the unconditioned, he nevertheless holds that their being is in a form asserted by consciousness."—*Christian Advocate.*

"The literary world has seen but few such authors as Herbert Spencer. There have been metaphysical writers in the same exalted sphere who before him have attempted to reduce the laws of nature to a rational system. But in the highest realm of philosophical investigation he stands head and shoulders above his predecessors; not perhaps purely by force of superior intellect, but partly owing to the greater aid which the light of modern science has afforded him in the prosecution of his difficult task."—*Boston Bulletin.*

"Mr. Spencer is achieving an enviable distinction by his contributions to the country's literature; his system of philosophy is destined to become a work of no small renown. Its appearance at this time is an evidence that our people are not *all* absorbed in war and its tragic events."—*Ohio State Journal.*

"Mr. Spencer's works will undoubtedly receive in this country the attention they merit. There is a broad liberality of tone throughout which will recommend them to thinking, inquiring Americans. Whether, as is asserted, he has established a new system of philosophy, and if so, whether that system is better than all other systems, is yet to be decided; but that his bold and vigorous thought will add something valuable and permanent to human knowledge is undeniable."—*Utica Herald.*

"Herbert Spencer is the foremost among living thinkers. If less erudite than Hamilton, he is quite as original, and is more comprehensive and catholic than Mansel."—*Universalist.*

BIBLE TEACHINGS IN NATURE.

By the Rev. HUGH MacMILLAN

1 Vol., 12mo. Cloth. Price, $2.00.

From the N. Y. Observer.

"These are truly original and delightful discourses, in which investigations of natural science are skilfully and often eloquently employed to establish divine revelation, and to illustrate its truths."

From the Hartford Morning Post.

"This is a work of rare merit in its way, and may be read with great profit and interest by lovers of Nature—by those who have the gift of *insight*, and who can look up 'through Nature to Nature's God' and see the 'invisible power and Godhead in the things which He has made.'"

From the Eastern Argus.

"The healthy mind delights in the beauties and mysteries of Nature, and this volume will be found both instructive and interesting."

From the Daily Enquirer.

"This is a beautifully written work, intended to make the studies of the Bible and of Nature doubly attractive, by pointing out the harmony which exists between them as revealed to the earnest students of both."

From the Norfolk County Journal.

"The author sees God everywhere revealed in the development of Nature,—finds Him in the works of pure and unobtrusive beauty; in the grand and impressive in scenery, and in the wonderful manifestations with which the world abounds."

ESSAYS:

MORAL POLITICAL, AND ESTHETIC.

In one Volume. Large 12mo.

CONTENTS:

I. The Philosophy of Style.
II. Over-Legislation.
III. Morals of Trade.
IV. Personal Beauty.
V. Representative Government.
VI. Prison-Ethics.
VII. Railway Morals and Railway Policy.
VIII. Gracefulness.
IX. State Tamperings with Money and Banks.
X. Reform; the Dangers and the Safeguards.

ALSO,

SOCIAL STATICS;

OR,

THE CONDITIONS ESSENTIAL TO HUMAN HAPPINESS SPECIFIED, AND THE FIRST OF THEM DEVELOPED.

In one Volume. Large 12mo.

All these works are rich in materials for forming intelligent opinions, even where we are unable to agree with those put forth by the author. Much may be learned from them in departments in which our common Educational system is very deficient. The active citizen may derive from them accurate systematized information concerning his highest duties to society, and the principles on which they are based. He may gain clearer notions of the value and bearing of evidence, and be better able to distinguish between facts and inferences. He may find common things suggestive of wiser thought —nay, we will venture to say of truer emotion—than before. By giving us fuller realizations of liberty and justice his writings will tend to increase our self-reliance in the great emergency of civilization to which we have been summoned.—*Atlantic Monthly*

HEAT,
CONSIDERED AS A MODE OF MOTION,

Being a Course of Twelve Lectures delivered before the Royal Institution of Great Britain.

BY JOHN TYNDALL, F. R. S.,
PROFESSOR OF NATURAL PHILOSOPHY IN THE ROYAL INSTITUTION—AUTHOR OF THE "GLACIERS OF THE ALPS," ETC.

With One Hundred Illustrations. 8vo, 480 pages. Price, $2.50.

From the American Journal of Science.—With all the skill which has made Faraday the master of experimental science in Great Britain, Professor Tyndall enjoys the advantage of a superior general culture, and is thus enabled to set forth his philosophy with all the graces of eloquence and the finish of superior diction. With a simplicity, and absence of technicalities, which render his explanations lucid to unscientific minds, and at the same time a thoroughness and originality by which he instructs the most learned, he unfolds all the modern philosophy of heat. His work takes rank at once as a classic upon the subject.

New York Times.—Professor Tyndall's course of lectures on heat is one of the most beautiful illustrations of a mode of handling scientific subjects, which is comparatively new, and which promises the best results, both to science and to literature generally; we mean the treatment of subjects in a style at once *profound* and *popular*. The title of Professor Tyndall's work indicates the theory of heat held by him, and indeed the only one now held by scientific men—*it is a mode of motion*.

Boston Journal.—He exhibits the curious and beautiful workings of nature in a most delightful manner. Before the reader particles of water lock themselves or fly asunder with a movement regulated like a dance. They form themselves into liquid flowers with fine serrated petals, or into rosettes of frozen gauze; they bound upward in boiling fountains, or creep slowly onward in stupendous glaciers. Flames burst into music and sing, or cease to sing, as the experimenter pleases, and metals paint themselves upon a screen in dazzling hues as the painter touches his canvas.

New York Tribune.—The most original and important contribution that has yet been made to the theory and literature of thermotics.

Scientific American.—The work is written in a charming style, and is the most valuable contribution to scientific literature that has been published in many years. It is the most popular exposition of the dynamical theory of heat that has yet appeared. The old material theory of heat may be said to be defunct.

Louisville Democrat.—This is one of the most delightful scientific works we have ever met. The lectures are so full of life and spirit that we can almost imagine the lecturer before us, and see his brilliant experiments in every stage of their progress. The theory is so carefully and thoroughly explained that no one can fail to understand it. Such books as these create a love for science.

Independent.—Professor Tyndall's expositions and experiments are remarkably thoughtful, ingenious, clear, and convincing; portions of the book have almost the interest of a romance, so startling are the descriptions and elucidations.

D. Appleton & Co., New York, have now ready,

A NEW
CLASS-BOOK OF CHEMISTRY,

IN WHICH

THE LATEST FACTS AND PRINCIPLES OF THE SCIENCE ARE EXPLAINED AND APPLIED TO THE ARTS OF LIFE AND THE PHENOMENA OF NATURE.

A NEW EDITION,
ENTIRELY REWRITTEN AND MUCH ENLARGED.

WITH

Three Hundred and Ten Engravings.

By EDWARD L. YOUMANS, M.D.

12mo. 460 pages.

The special attention of Educators is solicited to this work, on the following grounds:

I. It brings up the science to the present date, incorporating the new discoveries, the corrected views and more comprehensive principles which have resulted from recent inquiry. Among these may be mentioned the discoveries in *Spectrum Analysis*, the doctrines of the *Conservation and Correlation of Forces*, the researches of Berthelot on the *Artificial Production of Organic Substances*, the interesting researches of Graham on the *Crystalloid* and *Colloid* condition of matter, with many other results of recent investigation not found in contemporary text-books.

II. Avoiding excess of technicalities, it presents the subject in a lucid, forcible, and attractive style.

III. It is profusely illustrated with cuts of objects, apparatus, and experiments, which enable the student to pursue the subject alone or in schools without apparatus.

IV. Directions for experimental operations are much condensed, and descriptions of unimportant chemical substances are made very brief, or altogether omitted, thus obtaining space to treat with unusual fulness the "chemistry of common life," and the later revelations of this beautiful science.

V. It presents just such a view of the leading principles and more important facts of the science as is demanded for the purposes of general education.

VI. The work is arranged upon a natural method, the topics being so presented as to unfold the true order of Nature's activities. Part I treats of the natural forces by which matter is transformed. Part II, of the application of these forces to the lower or mineral world. Part III, of the organic kingdom, which rises out of the preceding; while Part IV, or Physiological Chemistry, completes the scheme in the world of life.

VII. It presents the science not only as a *branch* but as a *means* of education—a valuable instrument of intellectual culture and discipline.

VIII. It gives a clear exposition of the origin and nature of scientific knowledge and the value of scientific studies for purposes of education.

☞ *A Specimen Copy for examination will be sent, post paid, on receipt of 62 cents.*

www.ingramcontent.com/pod-product-compliance
Lightning Source LLC
Chambersburg PA
CBHW020845160426
43192CB00007B/786